Renaissance

Scenes of Industrial Reconversion in the Nord-Pas-de-Calais Coalfield

The Lieven Gevaert Series is a major series of substantial and innovative books on photography. Launched in 2004, the Series takes into account the ubiquitous presence of photography within modern culture and, in particular, the visual arts. At the forefront of contemporary thinking on photography, the books offer new insights on the position of the photographic medium within art historical, theoretical, social and institutional contexts.

The Series is produced by the Lieven Gevaert Research Centre for Photography, Art and Visual Culture (www.lievengevaertcentre.be) and covers four types of approaches: publication of outstanding monographic studies, proceedings of international conferences, book length projects with artists, translations and republications of classic material.

The Lieven Gevaert Series is published by Leuven University Press, and distributed in North America by Cornell University Press.

Renaissance

Scenes of Industrial Reconversion in the Nord-Pas-de-Calais Coalfield

Jorge Ribalta

Leuven University Press

Contents

Acknowledgments

This project was initiated after the invitation of Pia Viewing in 2013, then director of the CRP/ Centre régional de la photographie Hauts-de-France (formerly Nord-Pas-de-Calais), who provided the institutional engagement and the conceptual framework. A lot of coordination was required for shooting in some many sites. The following people and institutions were essential: Muriel Enjalran, Fatima Mehach, Christophe Olivier, Angeline Nison and Anaïs Perrin at the CRP/; Marie Patou and Catherine O'Miel, Mission Bassin Minier, Oignies; Karine Sprimont and Virginie Debrabant, Centre Historique Minier Lewarde; Justine Flahaut, site 9/9bis, Oignies; Sandrine Ghislain, site Arenberg, Wallers; Aude Dobrakowski, Musée Histoire Naturelle, Lille; Valérie Chevalier, Louvre-Lens; Nelly Turlutte and Serge Barruis, Maison Syndicale, Lens; Olivier Hardy, Archives du Monde du Travail, Roubaix; Aurore Atmania, Université Populaire Mineurs du Monde and Faculté Jean Perrin; Christiane Bouvart, from the Bibliotheque Municipale du Le Cateau Cambresis, provided important historical documentation on the city; Carrie Pilto, Musée Matisse, Le Cateau Cambresis; Bert Hogenkamp. Special thanks to Isabelle Duquenne, Bibliothèque Municipale, Lille.

The book was initiated by Hilde Van Gelder at the panel discussion held at the CRP/ on December 5, 2015. Thanks also to the other speakers in the panel, Muriel Enjalran and Jacques Lemière.

Finally, thanks to the partners of the publication: Muriel Enjalran at CRP/, Alexander Streitberger and Hilde Van Gelder at the Lieven Gevaert Centre and Veerle De Laet at Leuven University Press, Iris Dressler and Hans D. Christ from the Württembergischer Kunstverein, Stuttgart, and Julián Rodríguez, from gallery Casa sin Fin, Madrid.

Renaissance

> "Le musée est la figure de proue d'un bassin minier en pleine renaissance"
> **Brochure Louvre-Lens, 2013.**

The Nord-Pas-de-Calais coal-mining region crosses the French-Belgian border and is part of the historical Northern European heavy industry core, which extends into the quite near German Ruhr area. The coal exploitation started in the 18th century, peaked in the mid-20th century and lasted until the 1980s' Mitterrand presidential mandate, when the French Northern region around Lille was designed to develop a "new economy" center for culture industry related activities, strategically located between Paris, London and Brussels, with the new business district Euralille as its symbolic center. After de-industrialization and the foundation of a mining museum in Lewarde, opened in 1984, a pioneer "écomusée" institution in France, the region has participated in industrial heritage policies and campaigns, following the model of the Ruhr industrial region. A few of the former mining sites have been preserved and they define a historical industrial region that was declared world heritage by UNESCO in 2012, simultaneously to some Belgian sites. The most recent last step in this process of transition from industry into cultural and leisure economies is the opening of Louvre-Lens, a Louvre museum branch in a former mining site in Lens, in December 2012.

All of the shooting for *Renaissance* was made in March to May 2014. The title comes from the publicity of the Louvre-Lens museum and means to be an interrogation on the propagandistic promises of the new cultural economies. Does industrial reconversion really mean a renaissance? The coincidence with the slogan used for the fourth edition of Lille 3000 festival in 2015 was unexpected and undesired. It's probably a symptom of inner capitalism. Lille 3000 is a classic "postmodern" culture policy initiative promoted by the municipality and the region following the 2004 European Cultural Capital model in order to reinforce Lille's aspirations.

This photographic series is a tour of the monuments, an observation of such historical industrial landscape in transition. It is organized in eight scenes, composed of a variable number of photographs each. The scenes are articulated both geographically and conceptually, each addressing some specific topic.

The first three scenes are historical excavations or "memory scenes". They attempt to visibilize the , how the current Nord-Pas-de-Calais mining region is part of a longer European history that traces back to the rise of the modern capitalist nation-state system in the 16th century. The next three scenes deal with processes of resignification of mining sites as historical monuments. The last two scenes are organized around the dialectics between labor and leisure.

From Charles V to Louis XIV

The coronation of Charles V as the King of Spain in 1516 was the beginning of a period of Habsburg hegemony in Europe, in permanent dispute with the French Valois dynasty. Part of the current French Nord-Pas-de-Calais region was then Flanders and thus under the Spanish Empire/Habsburg rule. In the 1550s, just before his abdication in Brussels in 1555, some of the last military campaigns of Charles V against the Valois (i.e. Saint-Quentin) took place in this formerly border region between Flanders and France.

The sequence starts at the Louvre-Lens, where a portrait of Charles V's banker, Anton Fugger, painted by Hans Maler around 1529, is on display in the Renaissance section of the permanent collection. The so-called "maison espagnole" in Valenciennes is one of the various remaining local Flemish wooden architecture examples in the city from late 15th century. It is currently the tourist information center. In Cambrai, some of the original doors of the military citadel still exist, like the Porte Royale. Its columns show the characteristic Renaissance military style inspired by ancient Roman Empire architecture adopted by Charles V. In 1529 the peace treaty between François I and Charles V, known as "Cambrai's peace" or "peace of the ladies" ("paix des dames"), because the two monarchies were represented by women (Margaret of Austria and Louise of Savoy, respectively) was signed in Cambrai and ended the second war between the two dynasties. The end of the Charles V European rule period is marked by the 1559 Le Cateau Cambrésis peace treaty, signed by Henry II of France and Charles V's son, Philip II of Spain. The treaty ended a sixty-year war period between Habsburgs and Valois. The palace where the Le Cateau Cambrésis peace treaty was signed by both kings' ambassadors no longer exists. It was located where the current Parc Fénelon is, the gardens of the Palais Fénelon, a 18th century building which is currently the home of the Musée Matisse. Matisse was born in Le Cateau Cambrésis and the Museum was constituted in 1951, after a big artwork donation to the city by the artist. The photographs show details of the garden with some of the Museum workers, taken during the lunch break.

The last photographs are from the Centre Historique Minier de Lewarde, and show details of the exhibition *Le bassin minier, une conséquence de la bataille de Denain de 1712*. This was an exhibition that traced the origins of the coal exploitation in the region at the time of the Spanish Succession War (1701–1714) motivated by the struggles between the Habsburgs and the Bourbons for the Spanish monarchy, a struggle finally won by the Bourbons, who hold the Spanish throne since then. One of the battlefields in this international war was the future Nord-Pas-de-Calais coalfield, conquered by France to the formerly Flanders territories under Spanish rule. The exhibition included a portrait of Louis XIV, the Sun King, the king of France during the Succession War.

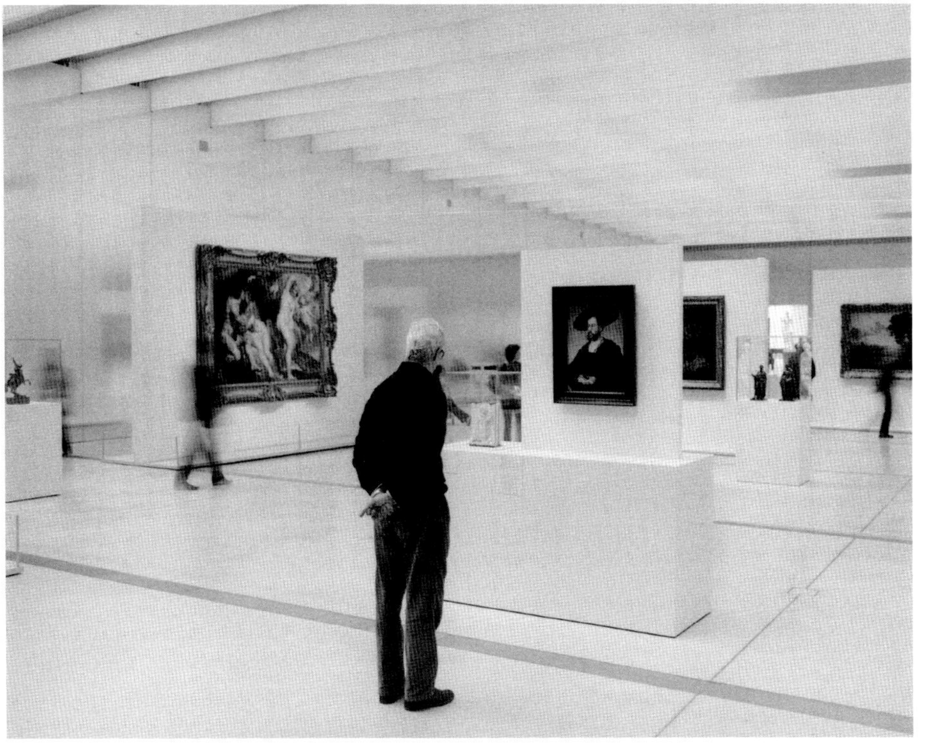

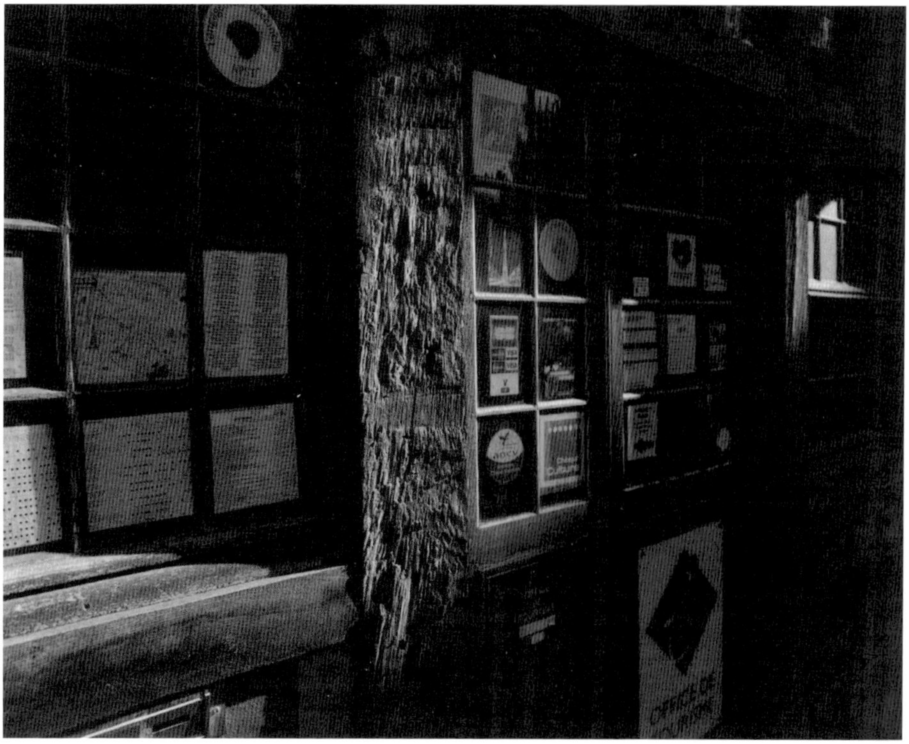

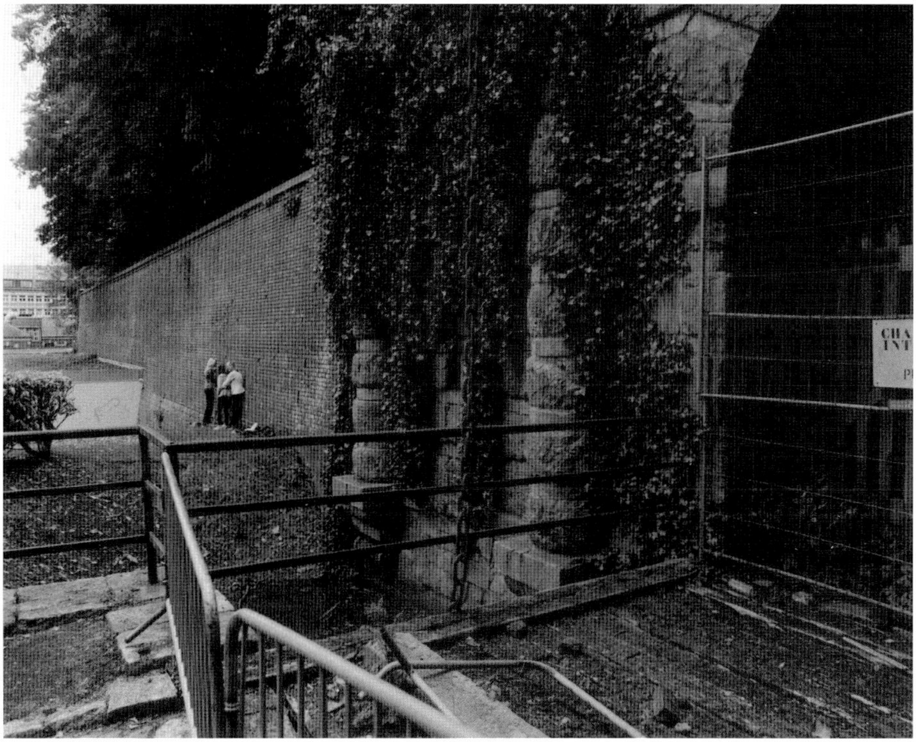

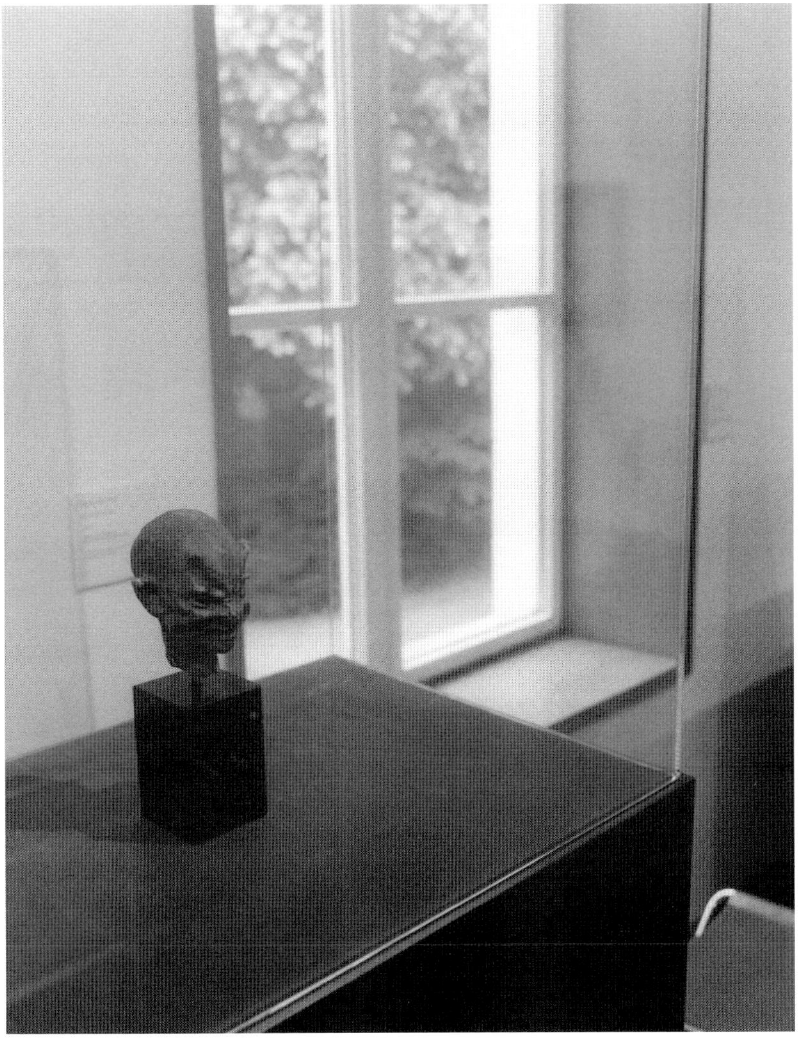

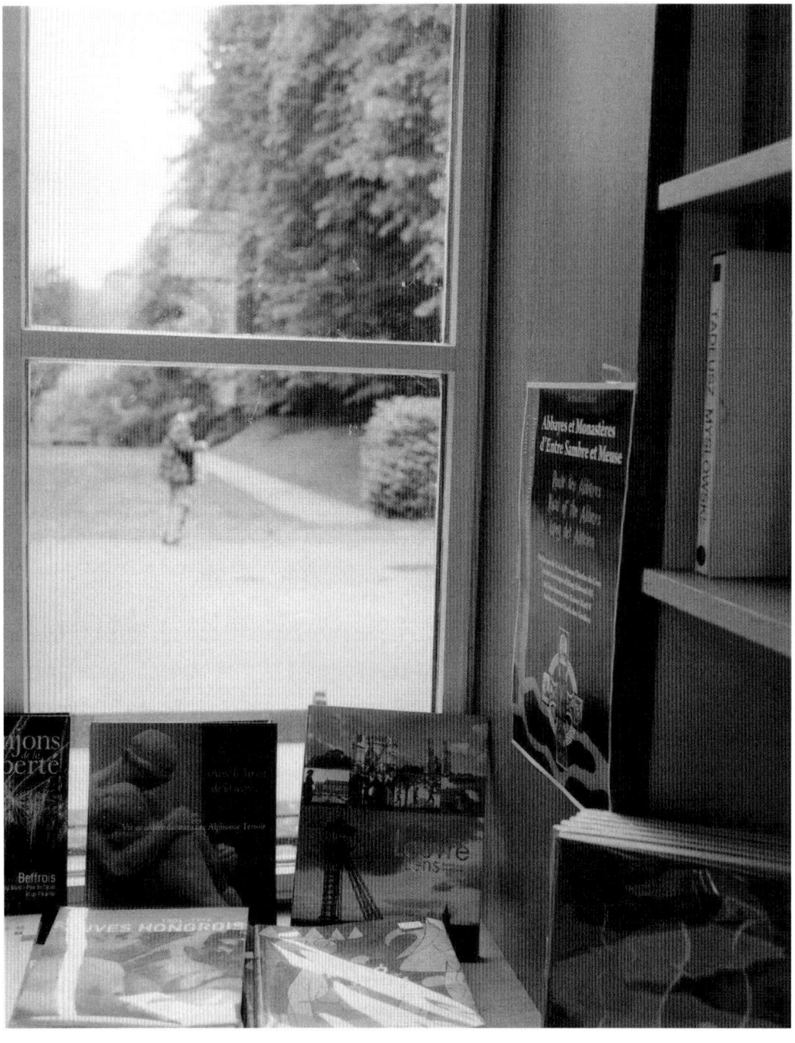
Abbayes et Monastères
d'Entre Sambre et Meuse
Beffrois

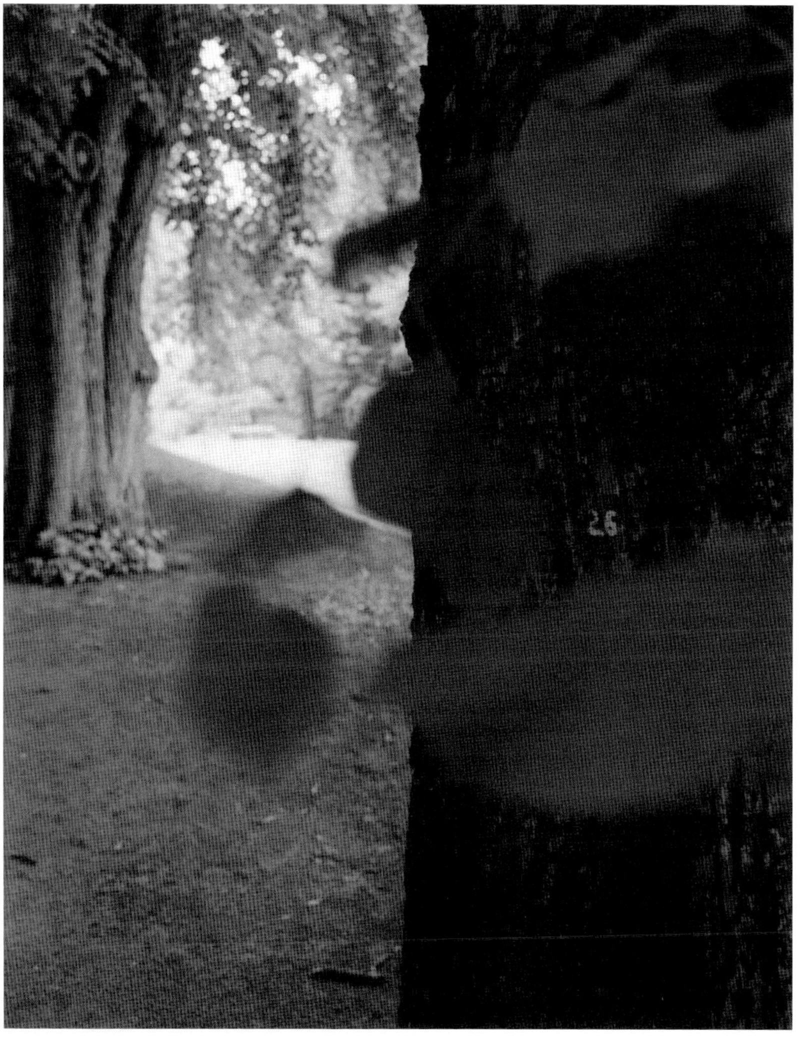

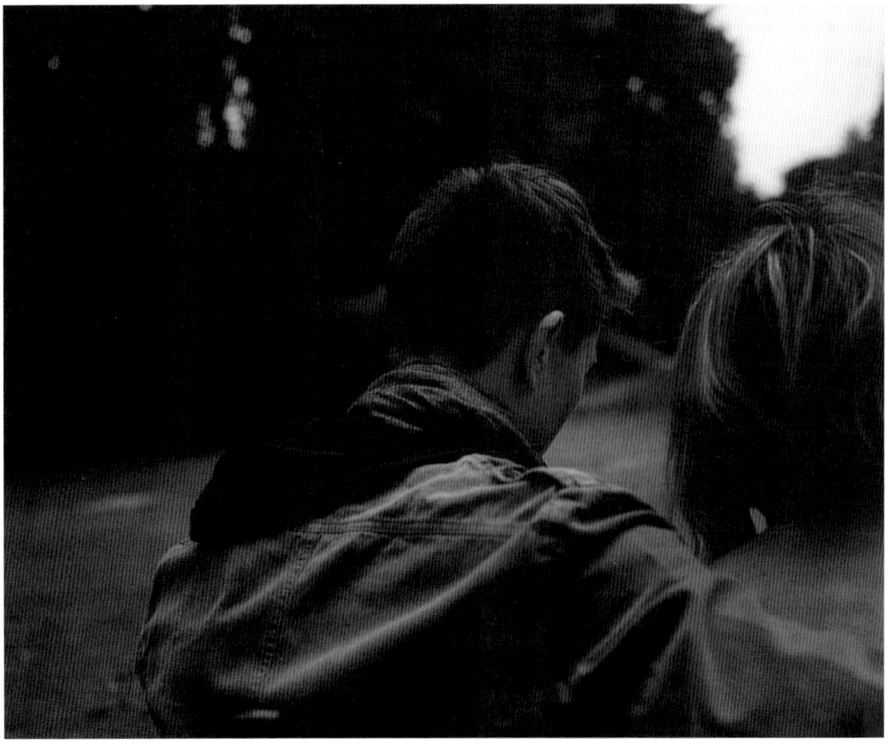

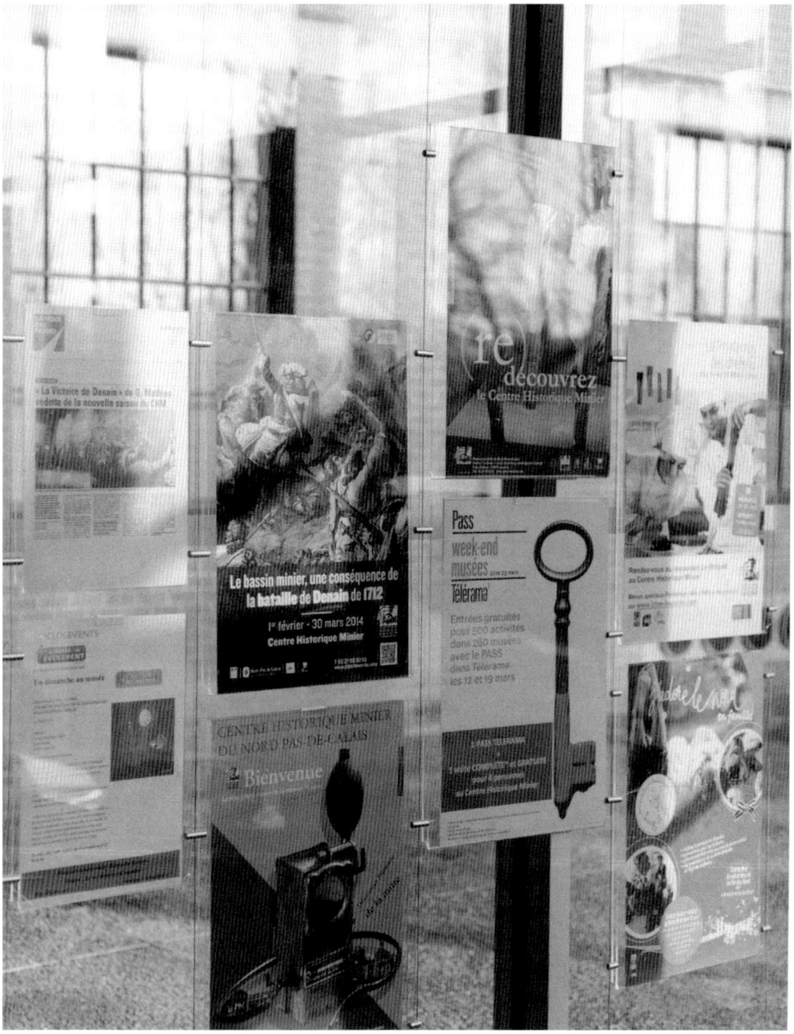
Le bassin minier, une conséquence de la bataille de Denain de 1712
1er février - 30 mars 2014
Centre Historique Minier
(re) découvrez le Centre Historique Minier
Pass week-end musées
Télérama
Entrées gratuites pour 500 activités dans 260 musées avec le PASS dans Télérama les 12 et 19 mars
CENTRE HISTORIQUE MINIER DU NORD PAS-DE-CALAIS
Bienvenue

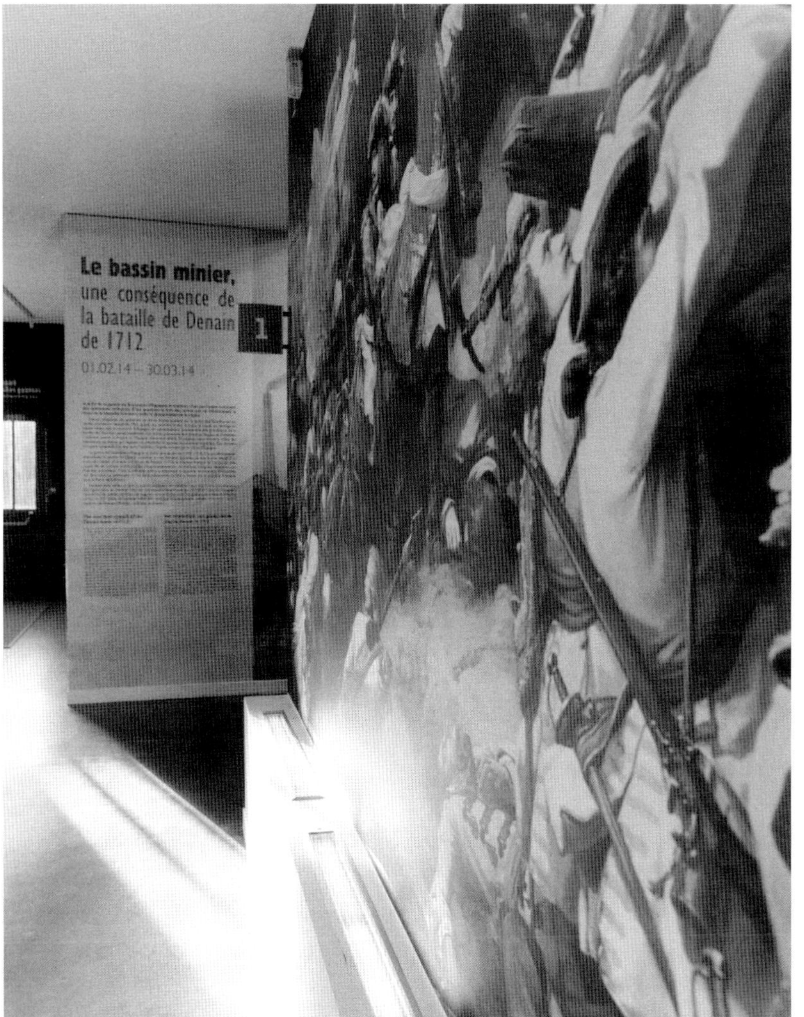
Le bassin minier,
une conséquence de
la bataille de Denain
de 1712
01.02.14 – 30.03.14
1

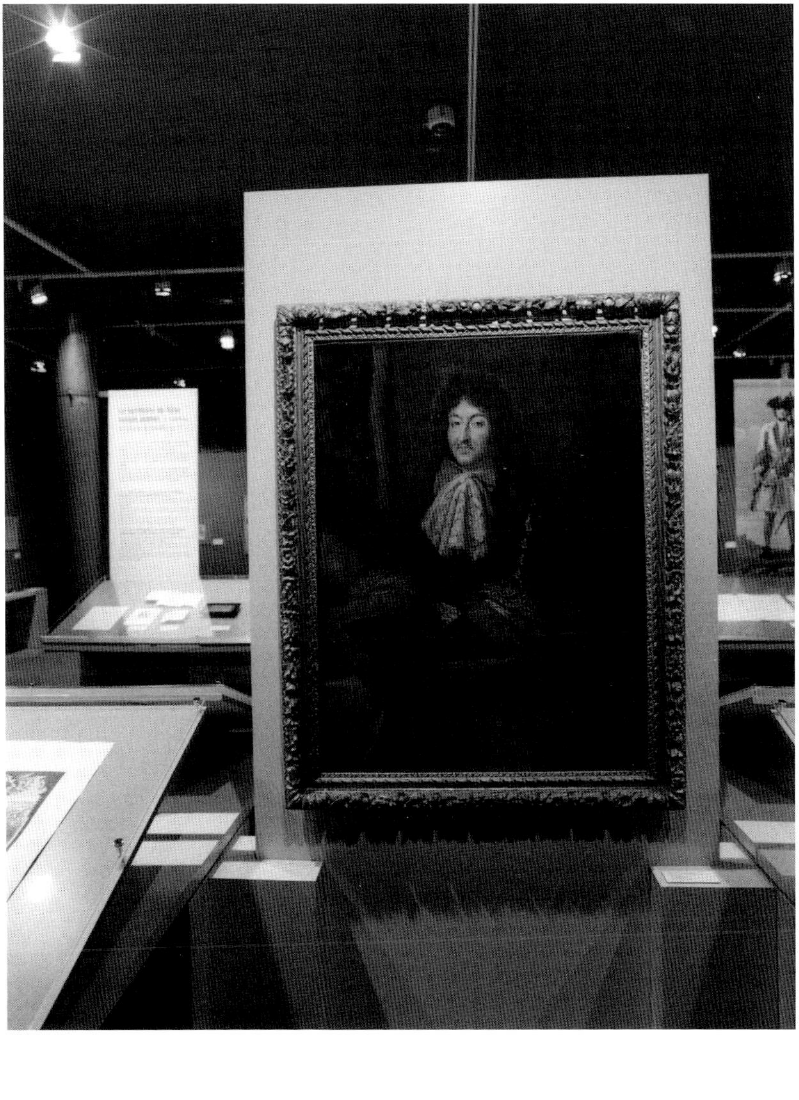

Small History of Photography

The second scene is an excavation in the history of photography, taken from some local remains and facts in the region.

In 1851, Louis-Désiré Blanquart-Évrard established his photographic printing enterprise in Loos, close to Lille. This represented the first technological revolution in the history of photography and the beginning of the era of photographic reproductibility and multiplicity. Blanquart-Évrard innovated printing technologies by introducing the albumen print and his *imprimerie* produced the first generation of photographic albums. But his enterprise had a short life and went into bankruptcy in 1855. He was also one of the local entrepreneurs who participated in the foundation of the Musée Industriel in Lille and he was one of the founding members of the Societé Française de Photographie in 1854. The sequence of photographs starts with a small tour through some remnants of Blanquart-Évrard's history in Lille. His legacy is today divided in two local institutions, the Bibliothéque Municipale, that holds the albums and prints and the Musée d'Histoire Naturelle that holds the negatives, archive and process materials, coming from the former collection of the Musée Industriel. In the photographs we can see the album *La Belgique* (1854), open by Place Royale and Parc de Bruxelles plates (which is the location of the historical Charles V palace in Brussels, his 1555 abdication site, whose remains still exist in the underground of the square and its surroundings), in the research room at the Bibliothéque Municipale; at the photographic reserve of the Musée d'Histoire Naturelle, one of the assistant curators holds Blanquart-Évrard's first album, the *Album de l'artiste et l'amateur* (1851) and his later book *La photographie. Ses origines, ses progrés, ses transformations* (1869). The site of the old Musée Industriel on 2 Rue Lombard is currently being renovated and transformed into a luxury apartment building. On the facade, the history of the building is described in an illustrated timeline, that starts with a reproduction of the painting *The moneylender and his wife* (1514) by the Flemish painter Quentin Matsys.

Blanquart-Évrard printing facilities or *maison imprimeri*e was in Loos, close to the Deule river. The precise location is today unknown due to changes in land registry over the last century and a half. The photographs are taken in the surroundings of the avenue Kuhlman and the rue departamentale 48, around the Deule docks, in the current administrative border between Lille and Loos, an industrial park.

The Centre régional de la photographie Hauts-de-France (formerly Nord-Pas-de-Calais) (CRP/) was founded in 1982 in Douchy-les-Mines, and installed in a former post office building. The rise of photographic institutions since the late 1970s and the 1980s in Europe was part of a structural and global transformation in photographic culture. The new social demands for cultural democratization after 1968 transformed the institutional landscape of culture. In the 1980s's Mitterand France, photography represented in a particular way a kind of social-democratic cultural policy medium whose emblem was the Mission Photographique de la DATAR (promoted by the French state for the documentation of the transformations of the national landscape during 1984 to 1988). The Mission DATAR was an echo of the Mission Héliographique from 1851, the first campaign in the history of photography, for the documentation of the historical monuments in France, which historically represents the rise of photography as an instrument for the contemporary nation-state ideology and its cultural self-legitimation. In the 1980s the CRP/ represented an impulse towards small-scale self-managed photographic education public spaces that would tend to progressively disappear or be reconfigured as art centers along the 1990s, due to the decline of institutional social-democracy in Europe and, most recently, the rise of neoliberal mass-tourism oriented cultural policies.

The photographs of the CRP/ show the old darkroom facilities and the archive with the collection. In the archive room, former CRP/ director Pia Viewing holds a demolition scene photograph from the exhibition *La Mine en œuvre 1890–1990*, a historical exhibition she initiated, researched and curated about the mining activity in the region from the photographic archives of the the Centre Historique Minier de Lewarde, presented at the CRP/ in 2010. It was a programmatic project for the CRP/ during Viewing's rule. The last image of this group is a portrait of three women, made behind the CRP/, at the place des Nations. The women are Fred Klaba, Marie Jo Noclain and Sylvie Dureuil, citizens of Douchy that participated in the project *Humaine*, by Marc Pataut made between 2009 and 2011 and presented in 2012 at the CRP/. Simultaneous to the opening of the Louvre-Lens and as its radical counterpart, Pataut's project was an attempt to problematicize modes of representation of the "people" and the institutional inscription of artistic activity aiming at experimentation and reinvention of the potential links between artistic methods and the activity of the social movements. This is a discursive space to which Pataut has contributed in a most significant and original way with his photographic activity over the last thirty years.

The last stop in this tour to the history of photography is Le Fresnoy in Tourcoing, in the Lille metropolitan area. Le Fresnoy is a flamboyant international artistic education experimental institution, supported by the French state that opened in the early 1990s. Its presence is part of the 1980s French national de-centralizing strategy for transforming Lille into a European cultural capital. Le Fresnoy is a high-tech building designed by architect Bernard Tschumi in a former working class ballroom. The photographs are made in the photographic facilities, the ultramodern darkrooms and film and print processing machinery, including sophisticated digital printing Lambda machines.

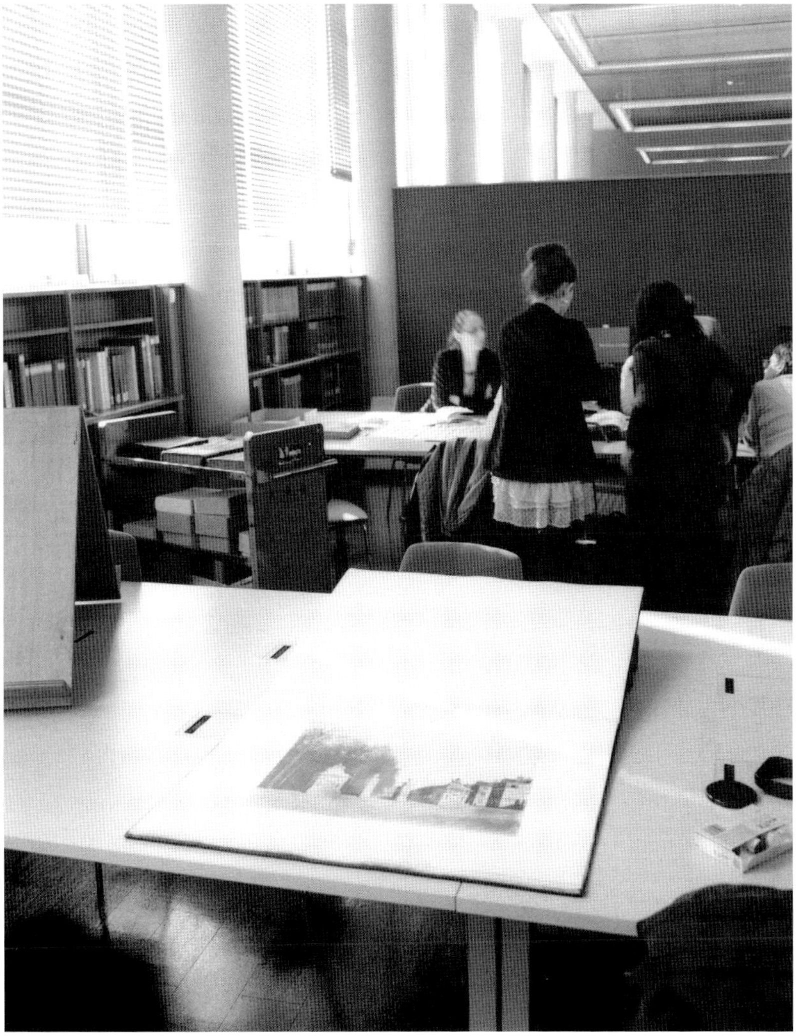

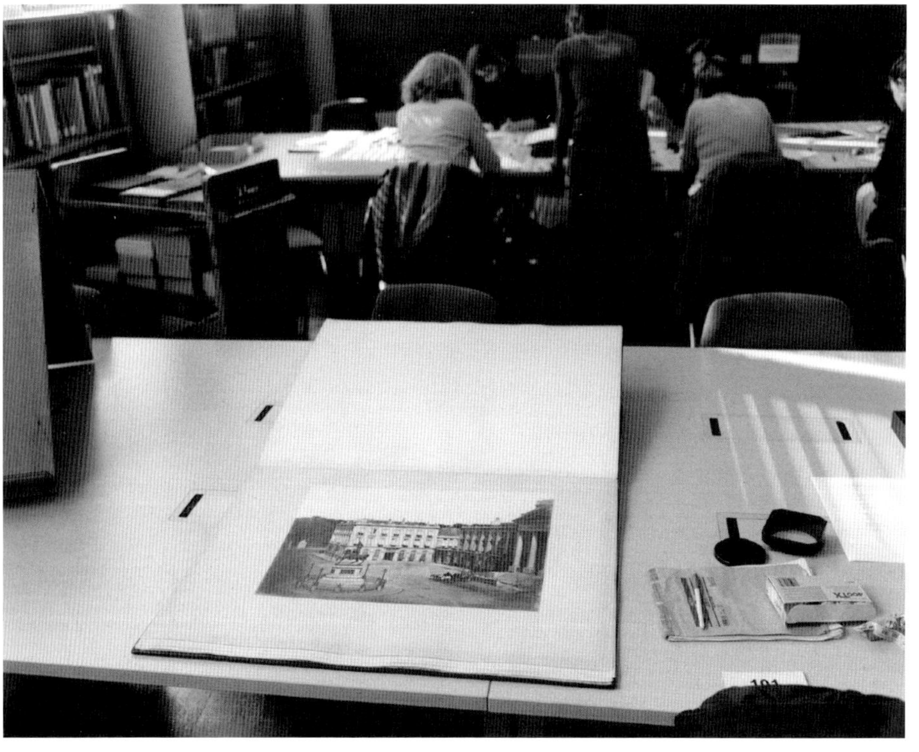

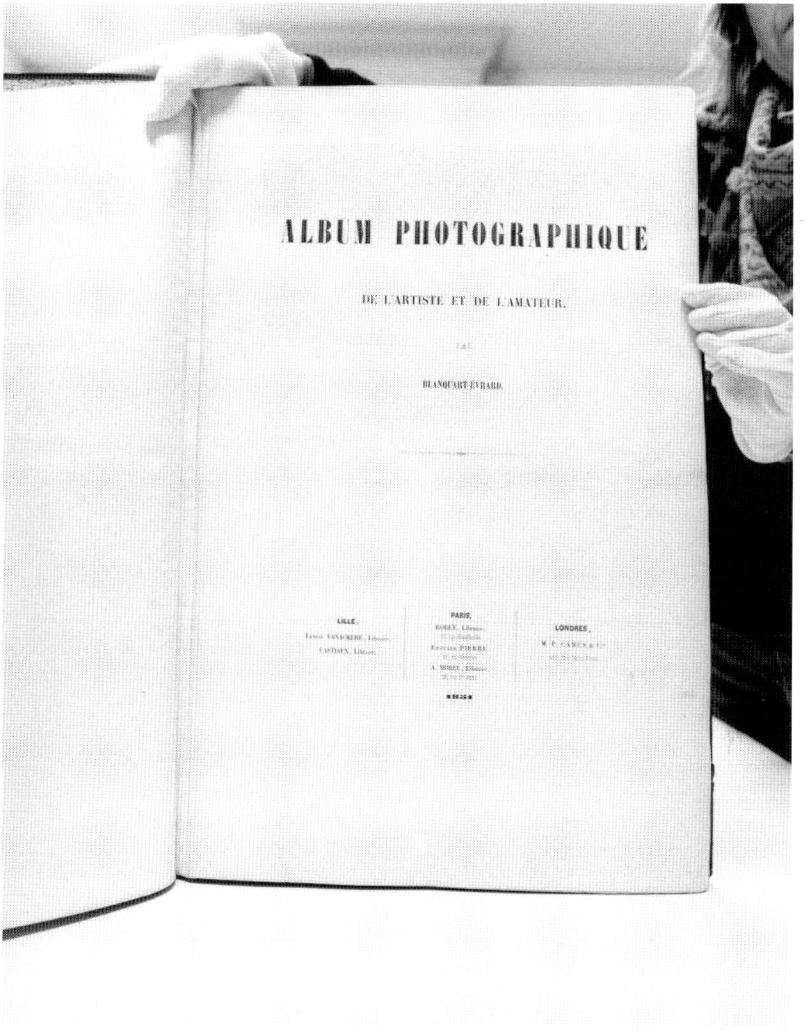
ALBUM PHOTOGRAPHIQUE
DE L'ARTISTE ET DE L'AMATEUR.
BLANQUART-EVRARD.
LILLE.
PARIS,
LONDRES.

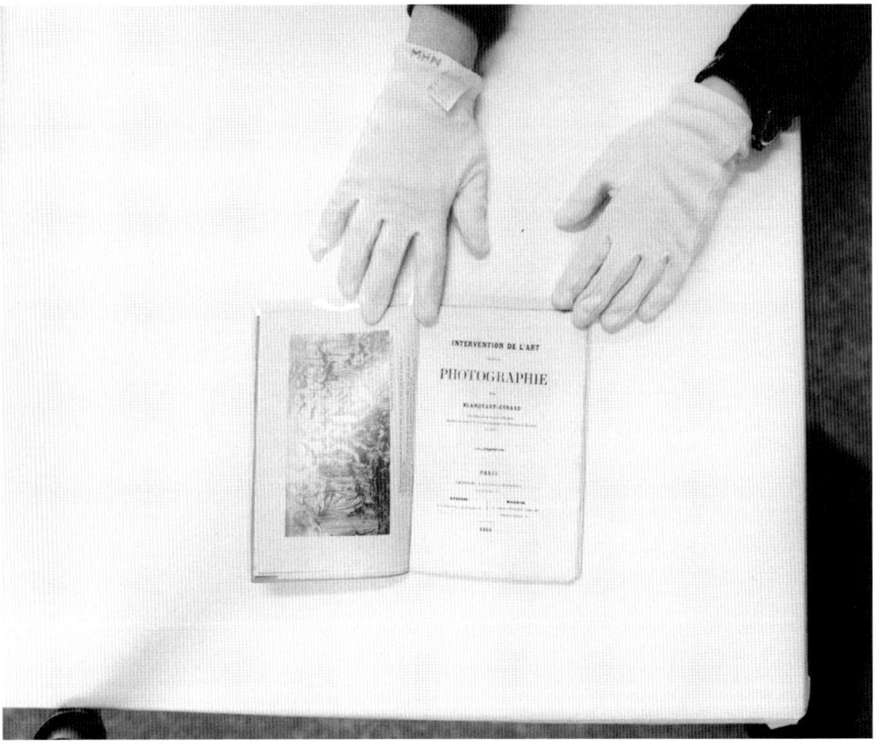
INTERVENTION DE L'ART
PHOTOGRAPHIE
BLANQUART-EVRARD
PARIS

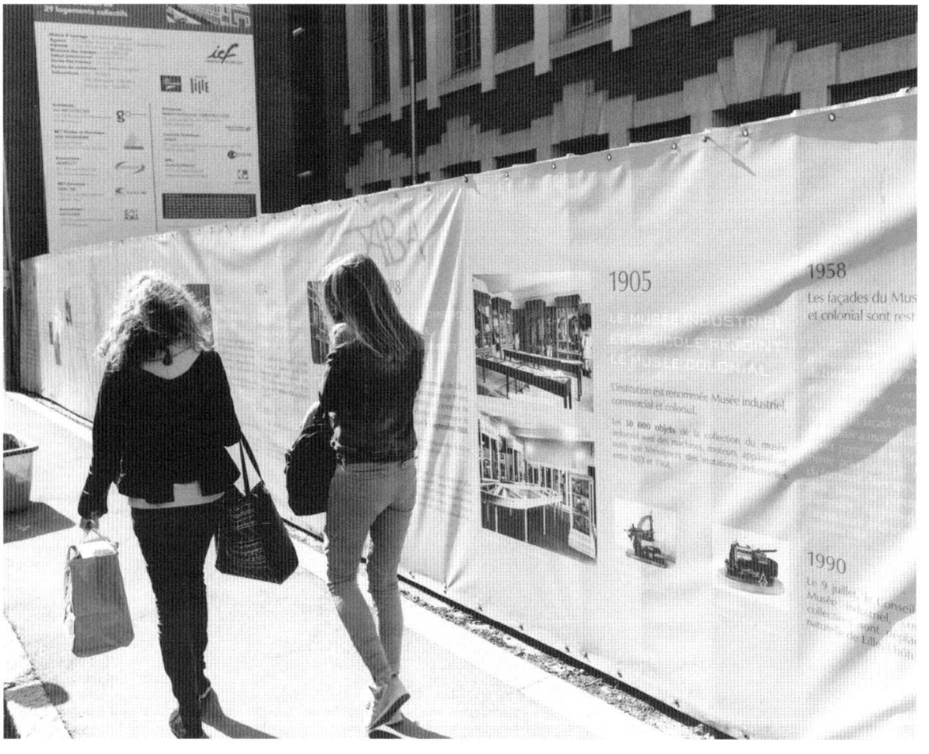
1905
L'institution est renommée Musée industriel, commercial et colonial.
1958
1990

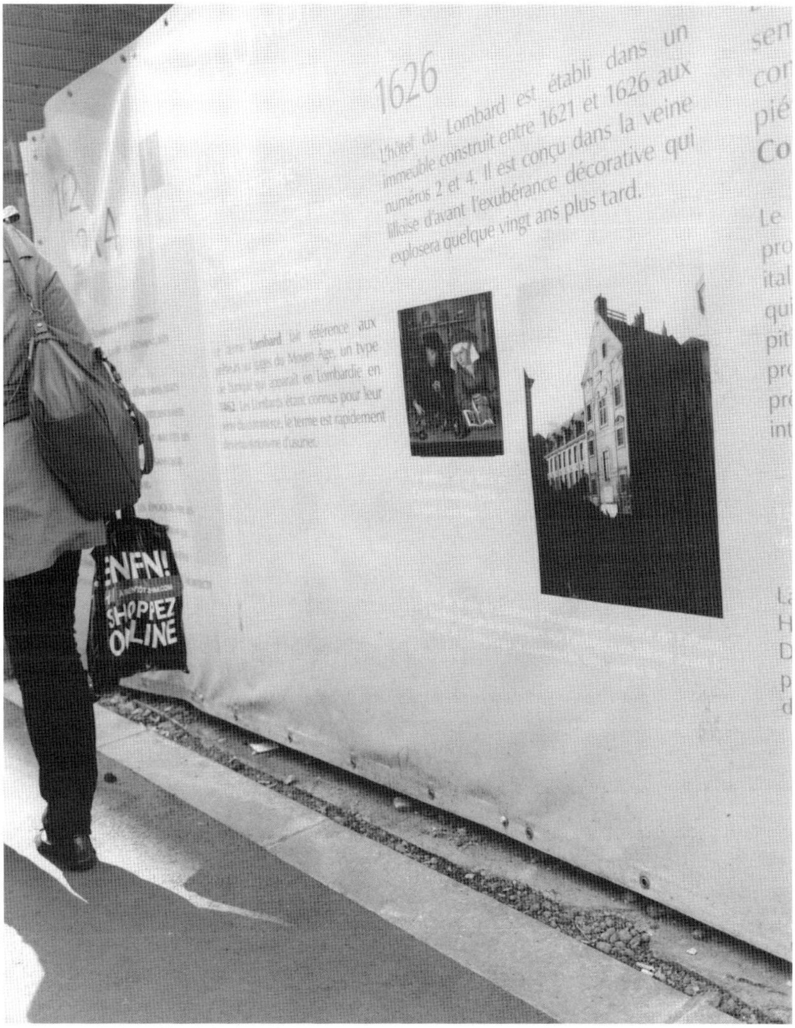
1626
L'hôtel du Lombard est établi dans un immeuble construit entre 1621 et 1626 aux numéros 2 et 4. Il est conçu dans la veine lilloise d'avant l'exubérance décorative qui explosera quelque vingt ans plus tard.
SHOPPEZ

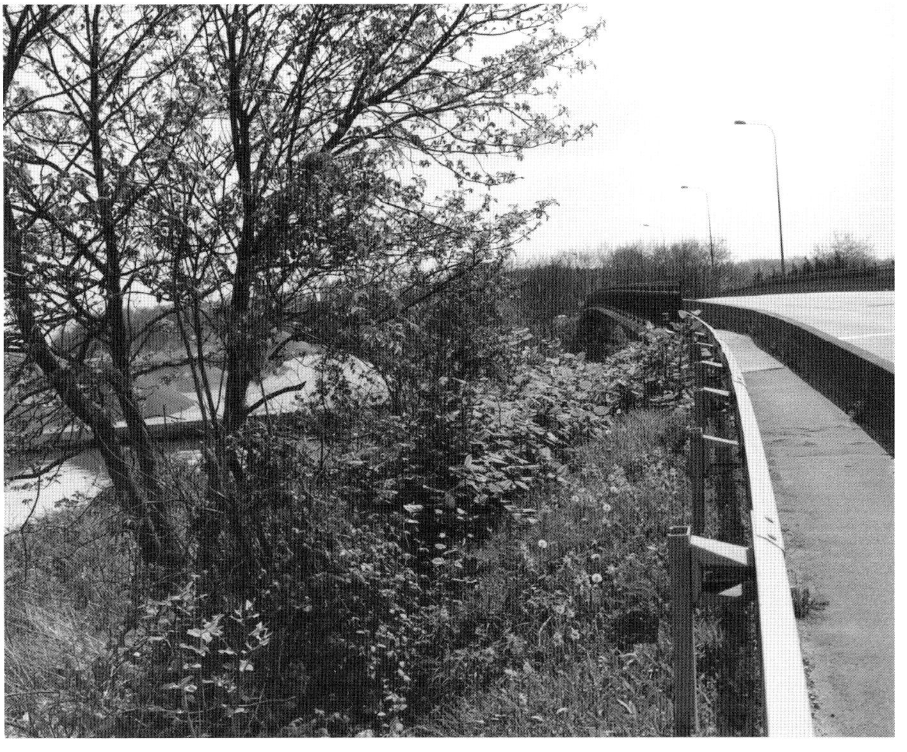

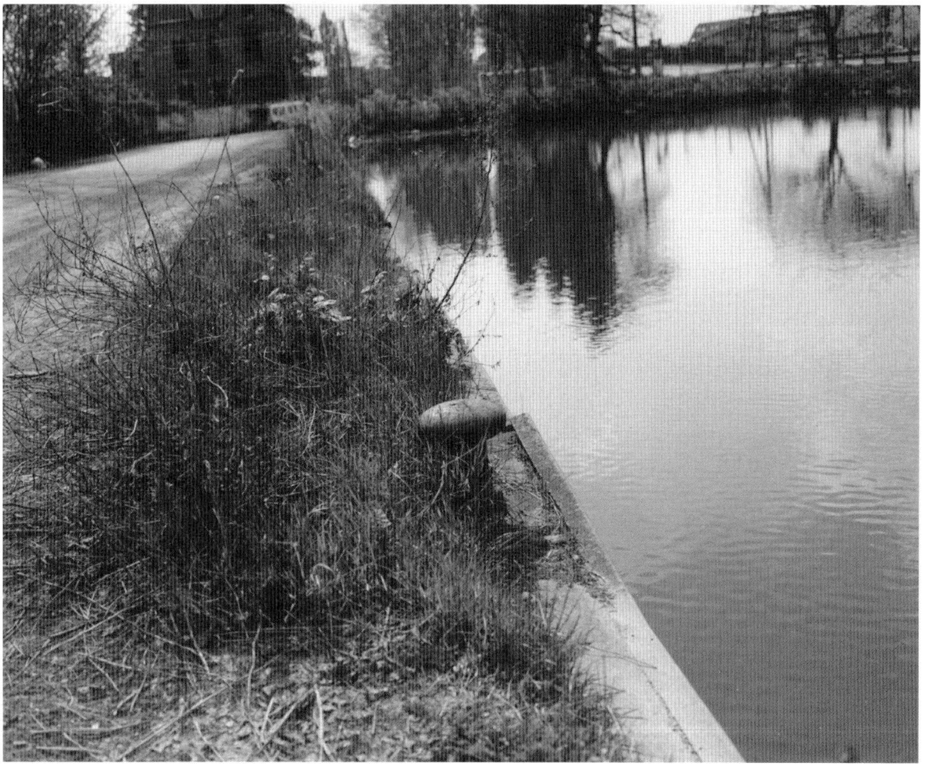

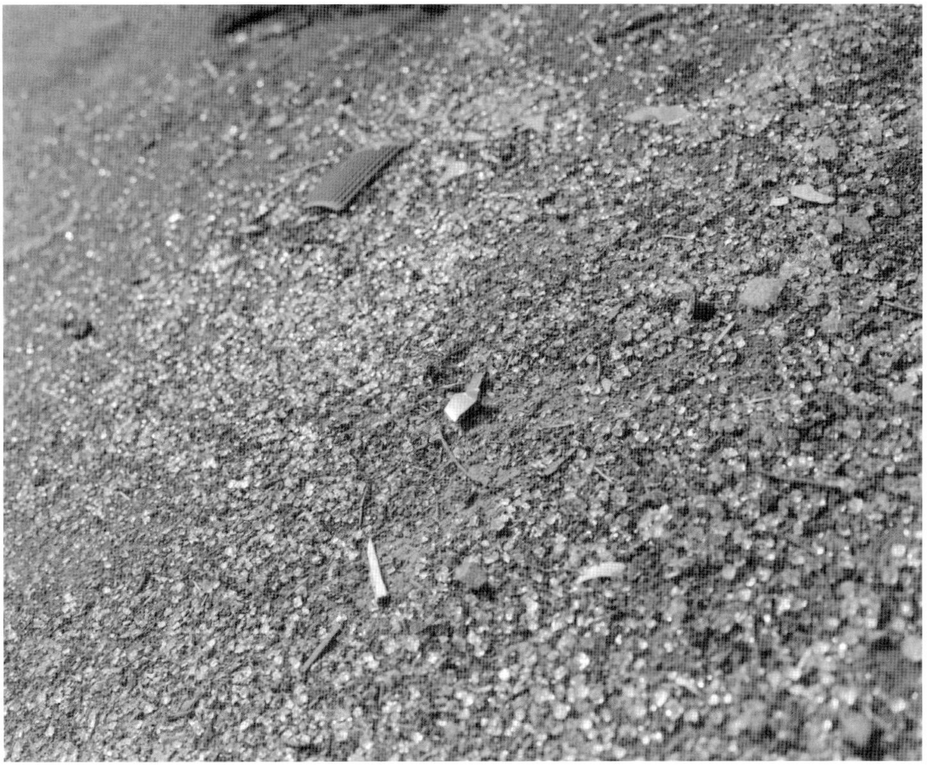

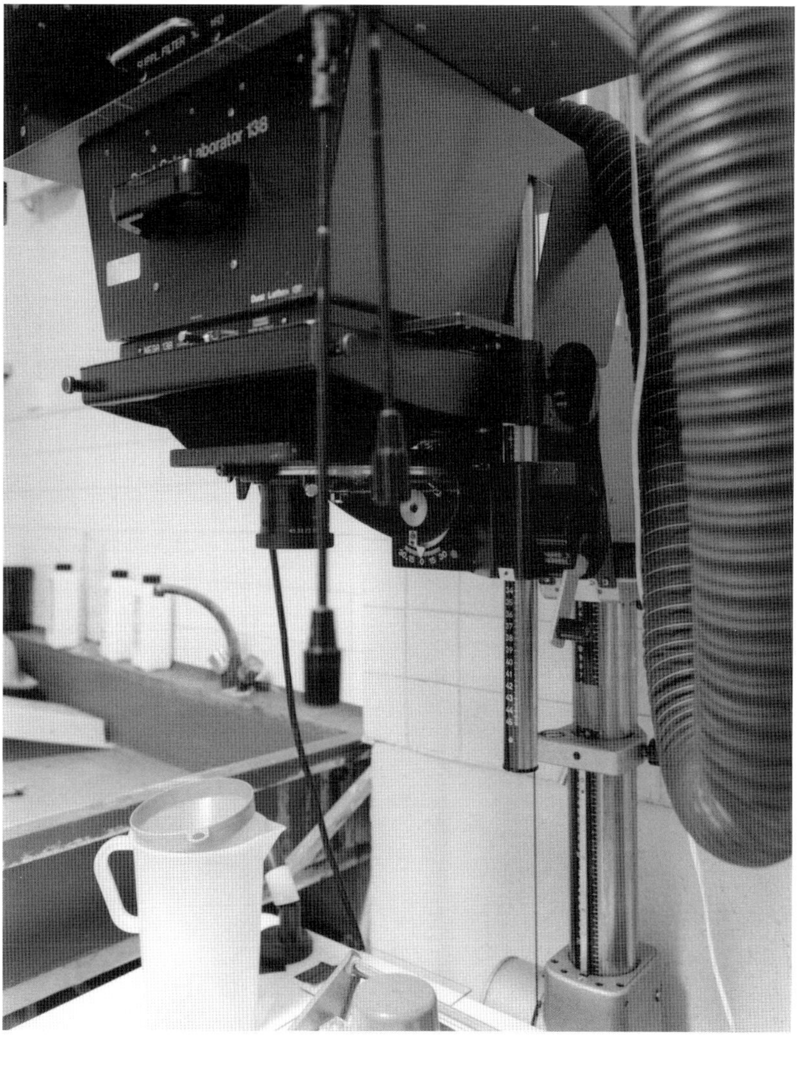

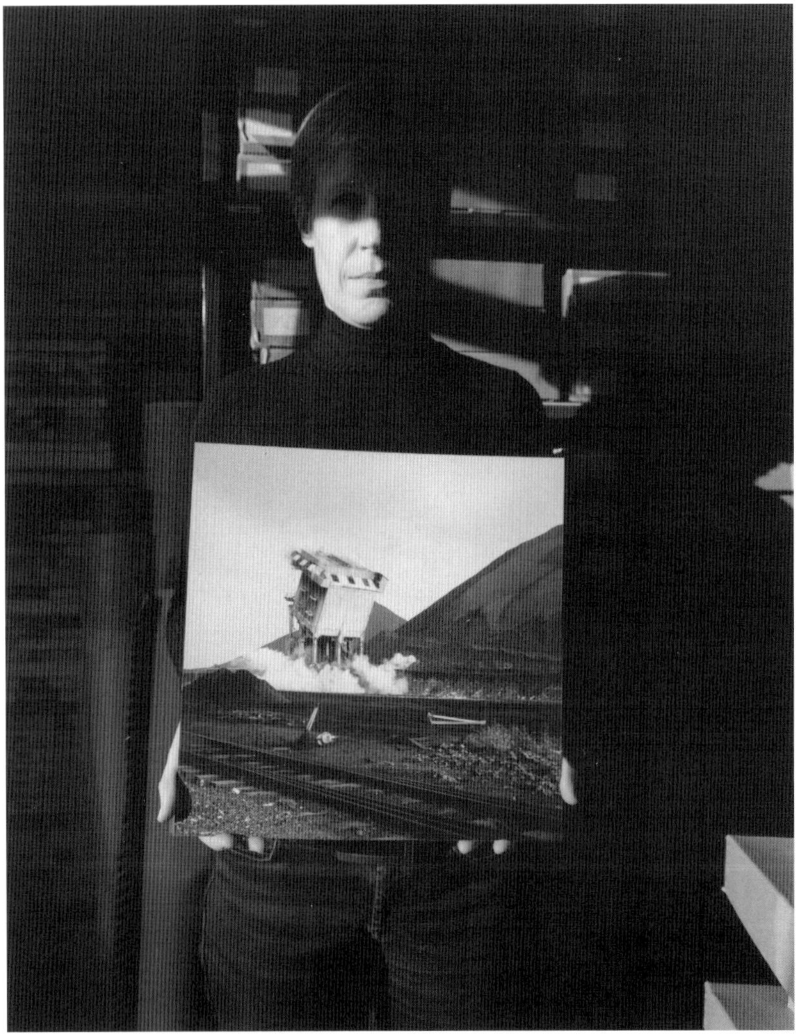

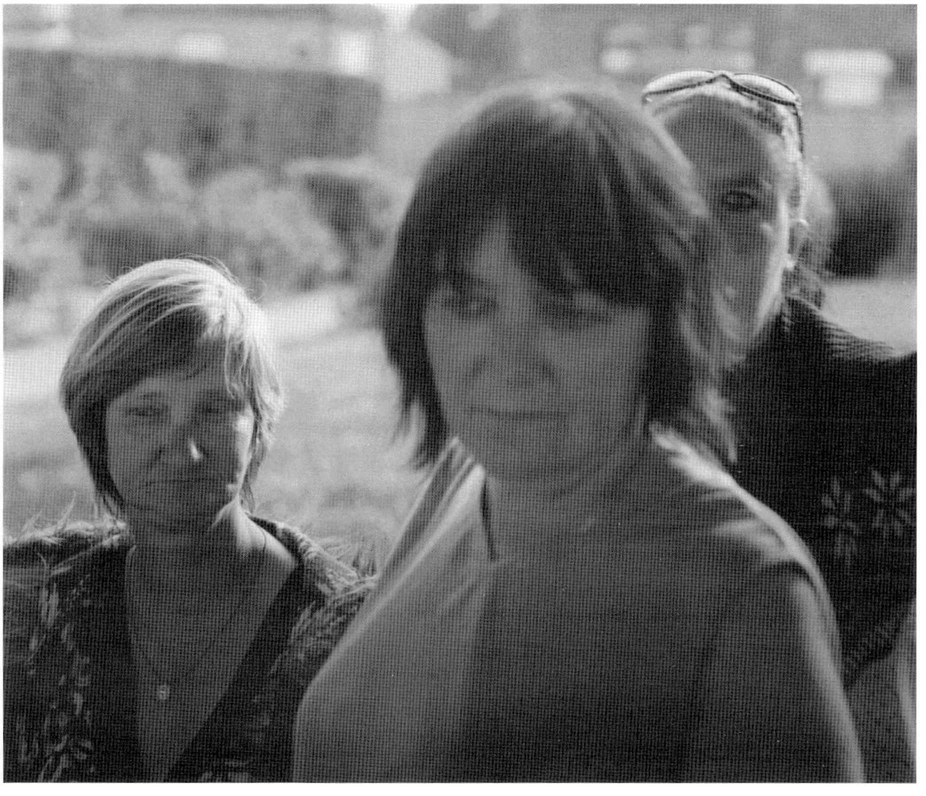

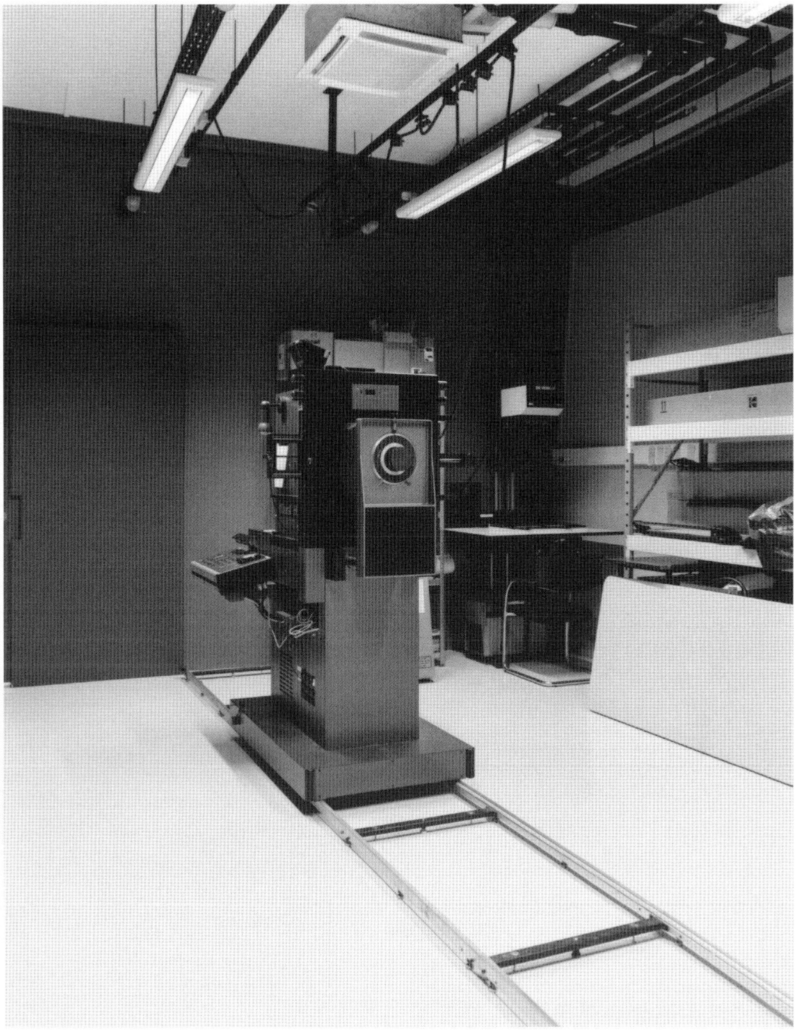

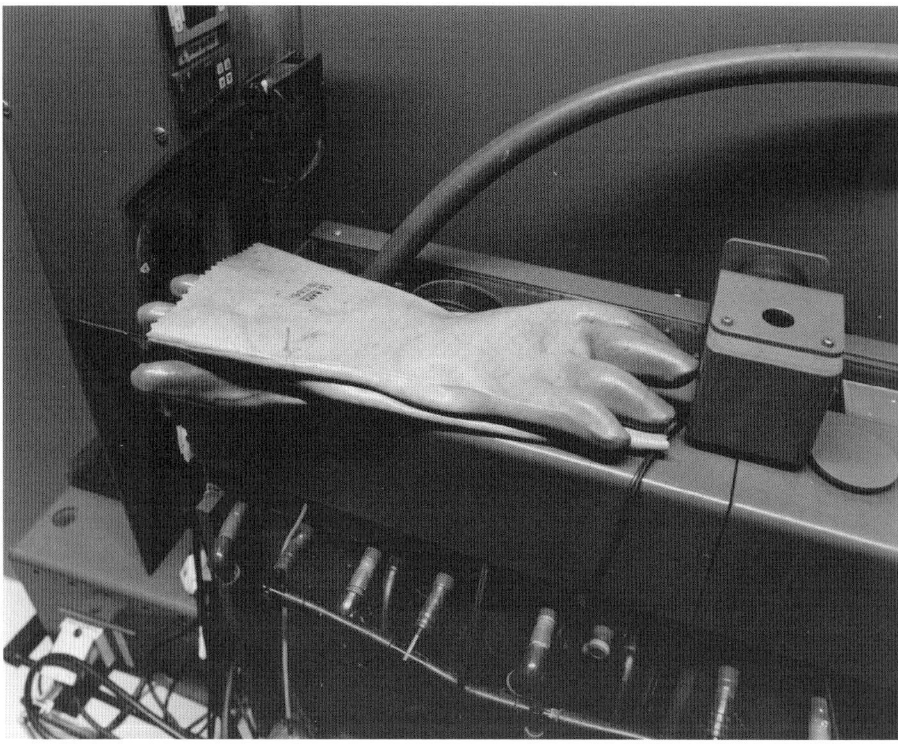

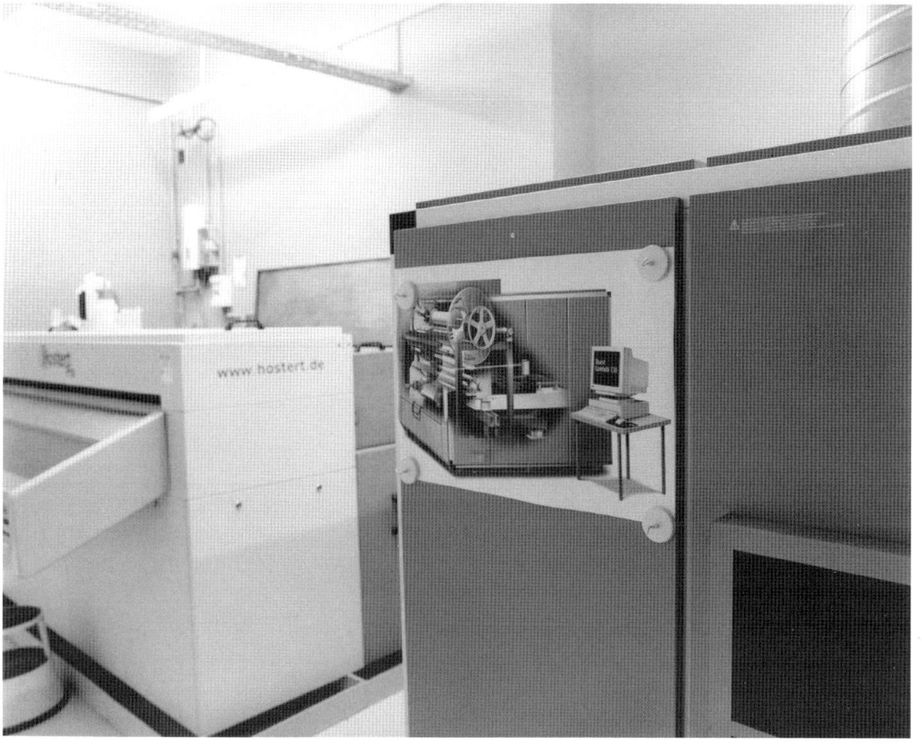
www.hostert.de

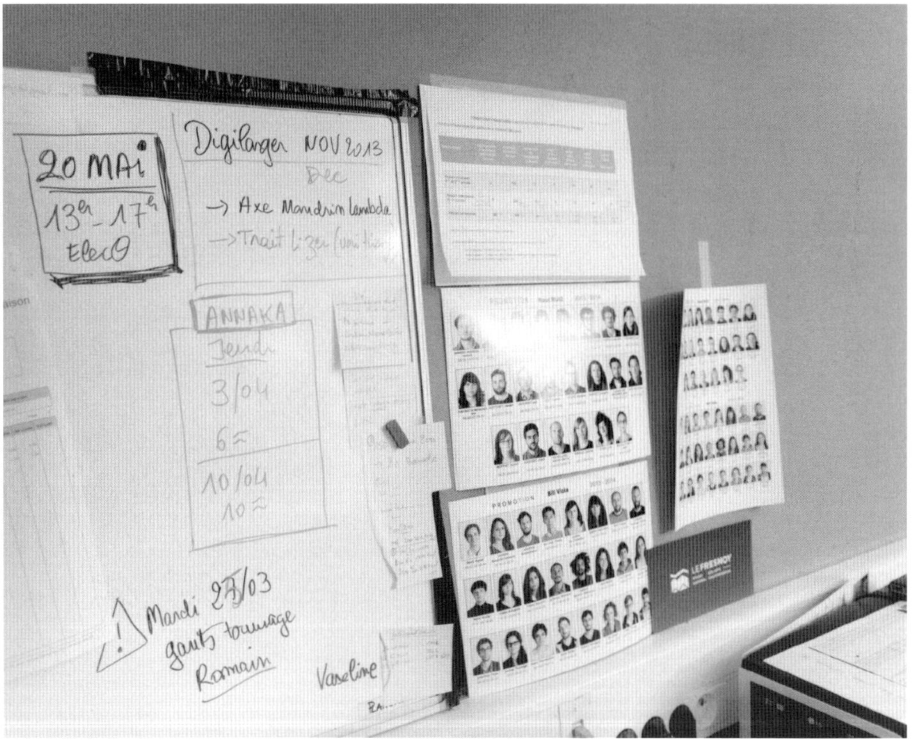
20 MAi
Digilanger NOV 2013
→ Axe Mandrin lambda
ANNAKA
Jeudi
3/04
10/04
Mardi
gants tournage
Romain
Vaseline
PROMOTION
LE FRESNOY

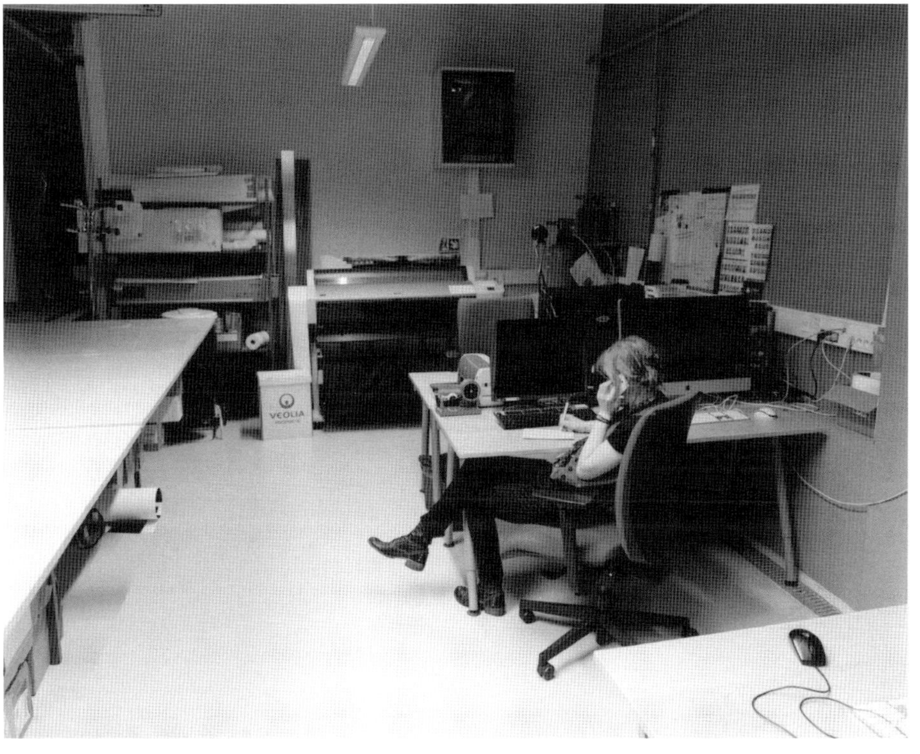
VEOLIA

Borinage

The third and last "memory" scene is a tour of the sites related to *Misère au Borinage* film (1933) shot by Joris Ivens and Henri Storck in the Mons area in Belgium, very close to the French border. The film is a classic of the 1930s golden age of documentary that denounced the poor living conditions of the coal mineworkers in the Borinage region in the Wallonie, which is part of the same coalfield of the Nord-Pas-de-Calais region. Ivens is one of the historic intellectual and artistic fathers of documentary in the modernist avant-garde period between the wars. He introduced the important concept and method of *mise en scéne documentaire*. The impulse for documentary in the late 1920s was the self-representation and political empowerment of the working class at the rise of mass-media democracy. Ivens was one of the founders of the Dutch Worker-Photography branch in Amsterdam in 1930 and one of the key artists involved in such Movement in the 1930s. My own photographic observation is grounded on the study and recognition of such genuine modernist documentary impulse.

The tour includes Hornu, Wasmes, Quaregnon and the Levant de Mons in Bray, the esplanade where the Roi Albert barracks were installed at the time of the film. Even if the extreme poverty of 1930s has luckily disappeared, most of the places seem to remain as not particularly prosperous working class areas. Some of the photographs show campaign posters related to the European Elections of Sunday May 25, 2014.

The last photographs correspond to Le Grand Hornu site in Hornu, which is a beautiful neo-classic former mining industrial site and housing complex, declared world heritage by UNESCO in 2012, simultaneously to the Nord-Pas-de-Calais sites declaration. The mining activity ceased in the 1960s and in the 1980s it was reconverted into a cultural space. At the moment it is the home of two institutions: the MAC's, the contemporary arts museum of the Federation Wallonie-Bruxelles, opened in 2002 and Grand-Hornu Images (GHI), a design and crafts exhibition venue, managed by an homonymous association. The site and its various institutions receive public funding from various public and private sources, from local, regional, state, European and even international scales, like the UNESCO. The photographs show the surroundings of the complex, the museum exhibition spaces and facilities, including the shop and the cafeteria, and some portraits of workers at the MAC's, specifically the "guide conferencier" Sebastien Laurent and the guard Patrizia Notarrigo. Are the current young and university trained under-paid, precarious education and mediation workers of art institutions the equivalent to 1930s' industrial workers' proletarian faces?

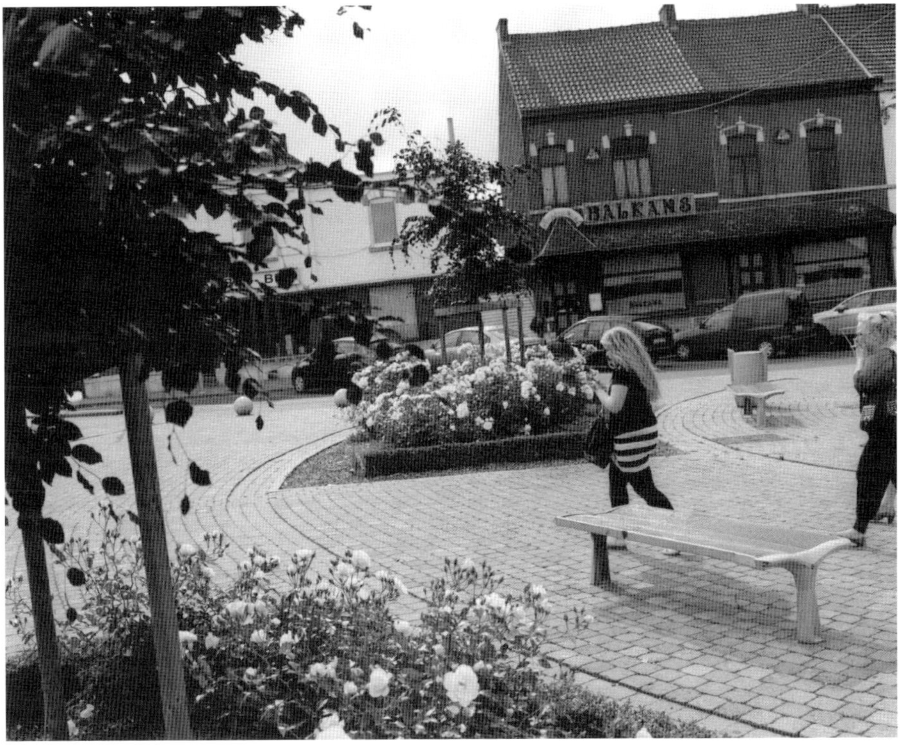
BALKANS

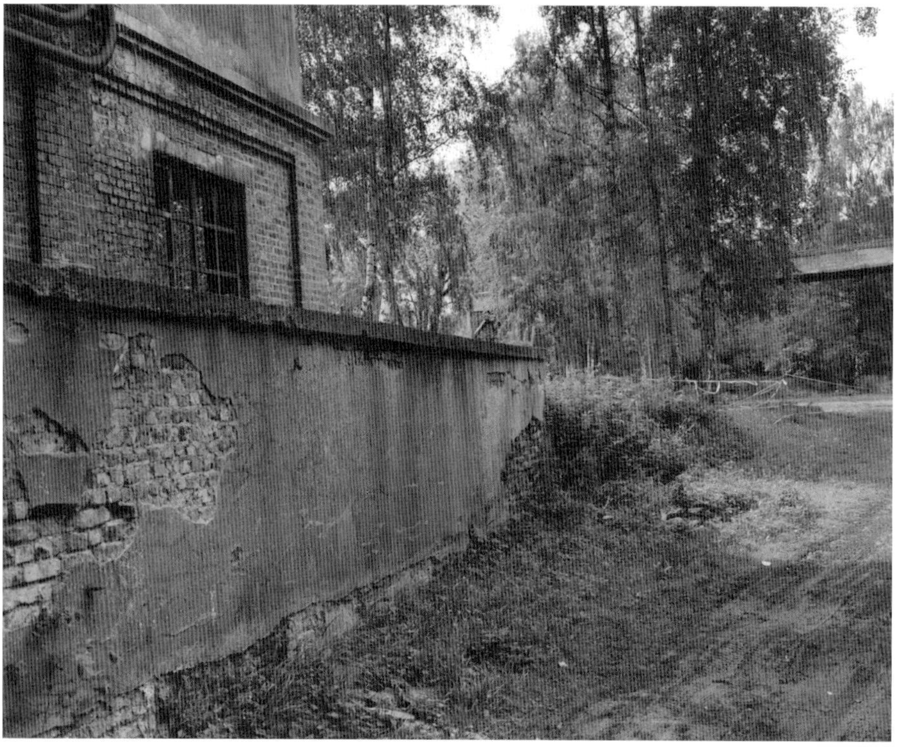

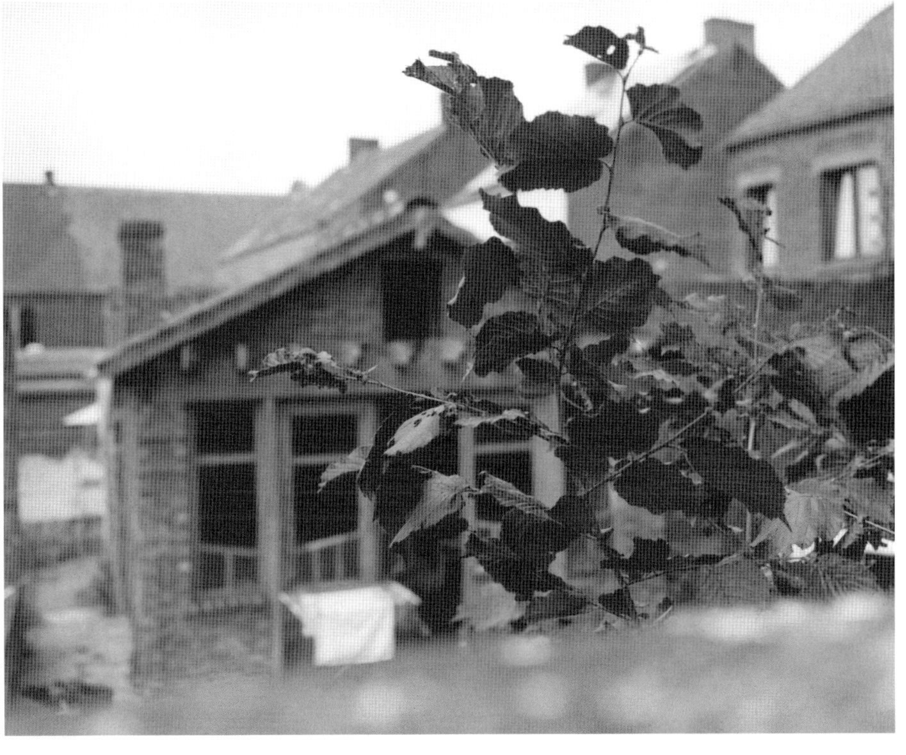

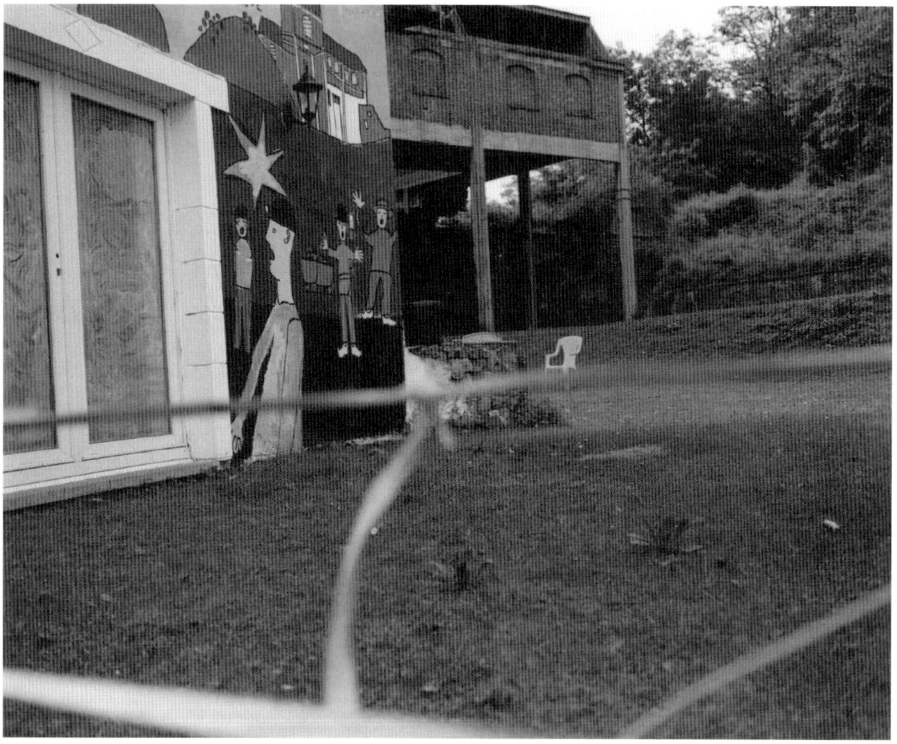

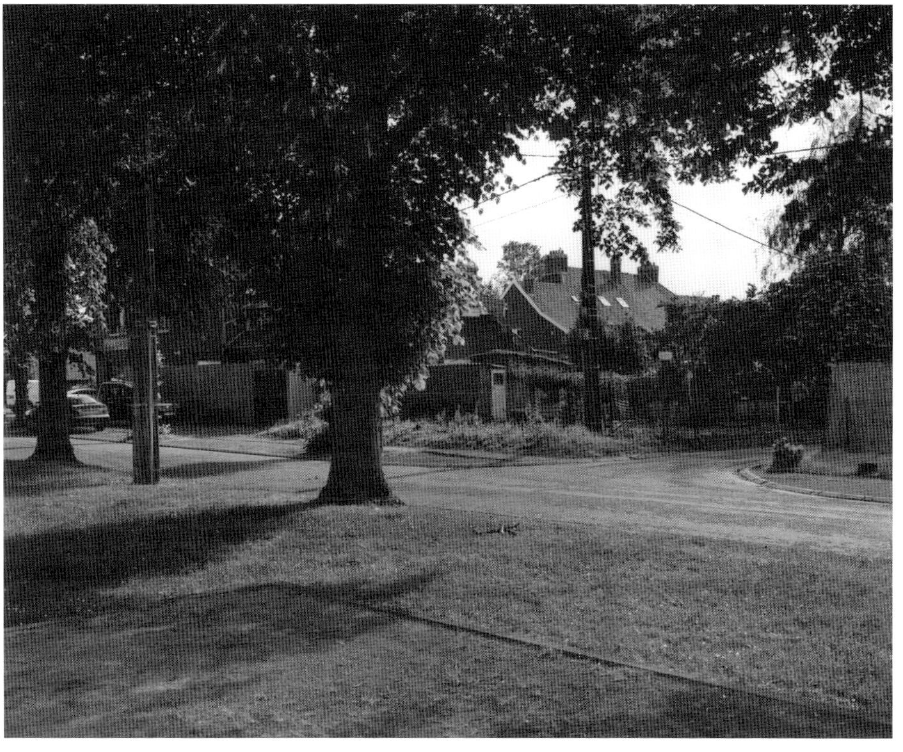

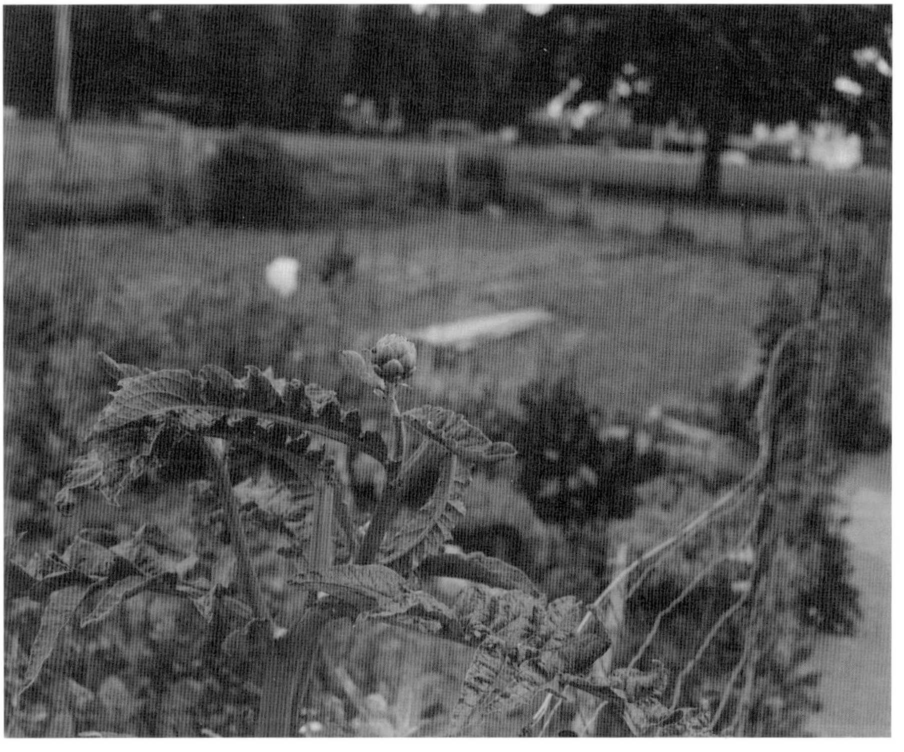

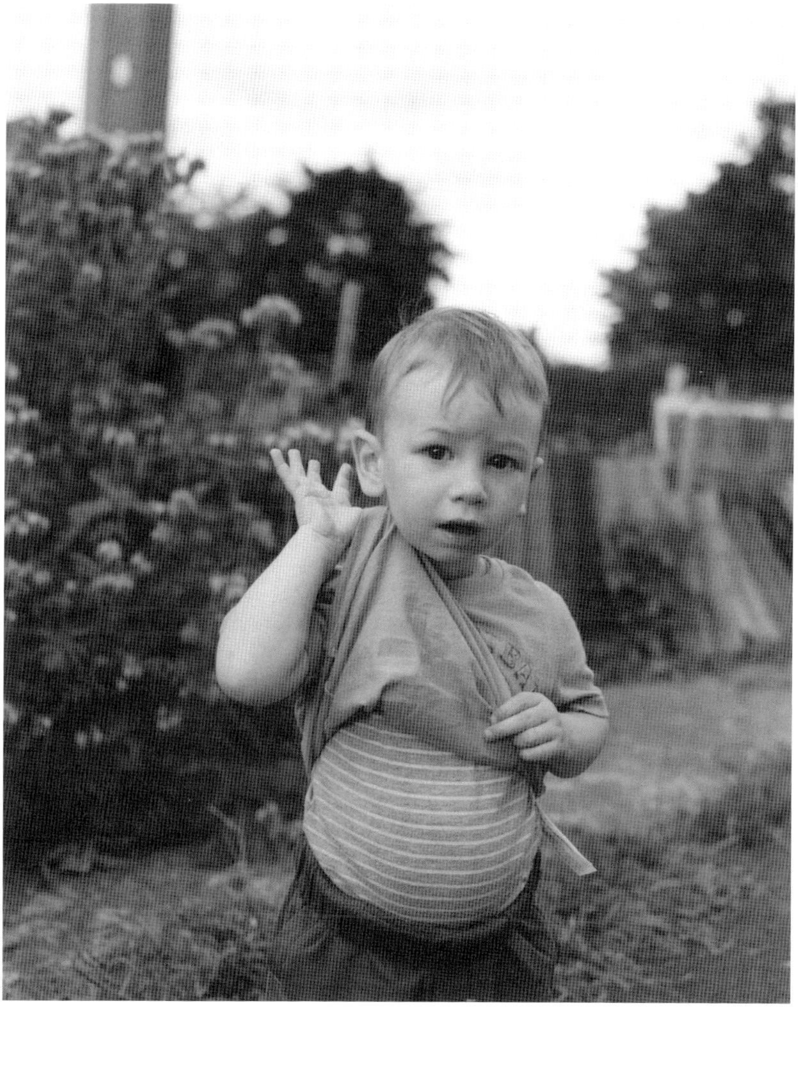

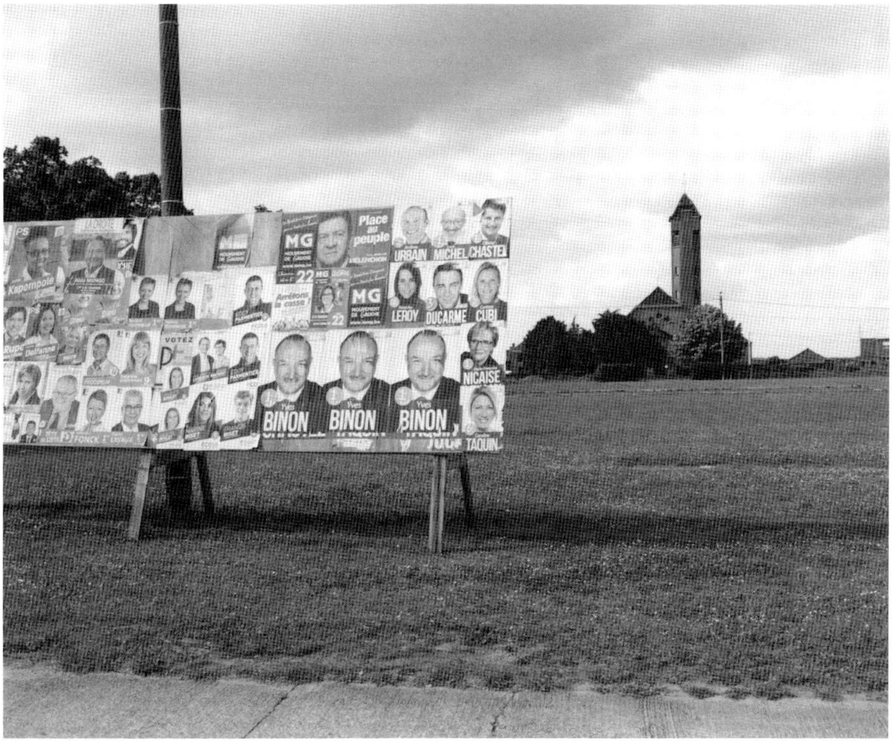
MG
Place au peuple
URBAIN
MICHEL
CHASTEL
LEROY
DUCARME
CUBI
MG
22
VOTEZ
NICAISE
Yves
BINON
Yves
BINON
Yves
BINON
FONCK
TAQUIN

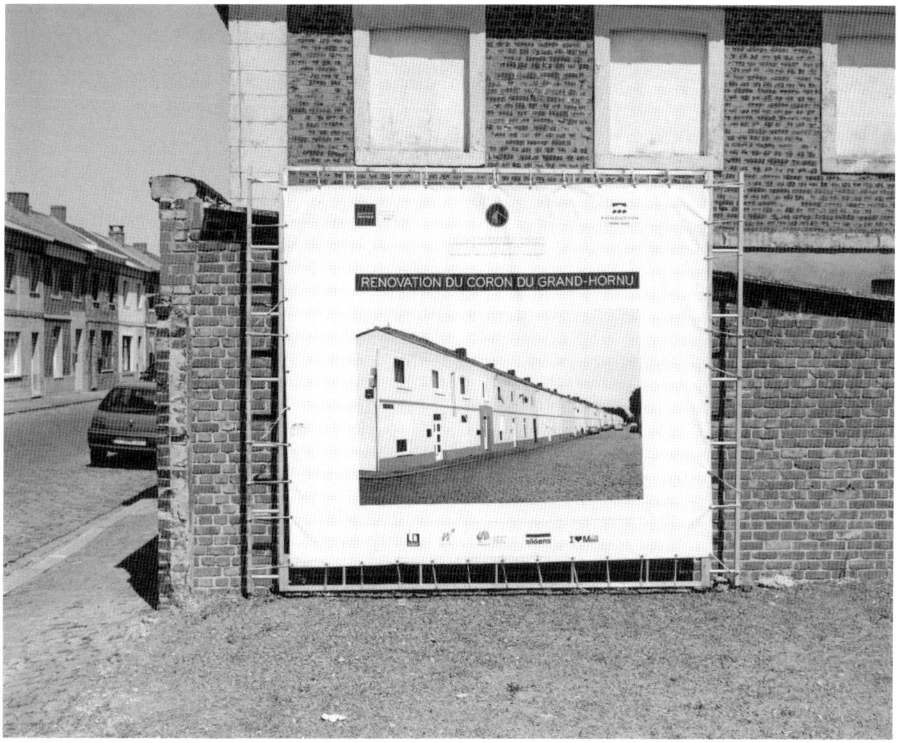
RENOVATION DU CORON DU GRAND-HORNU
sikkens

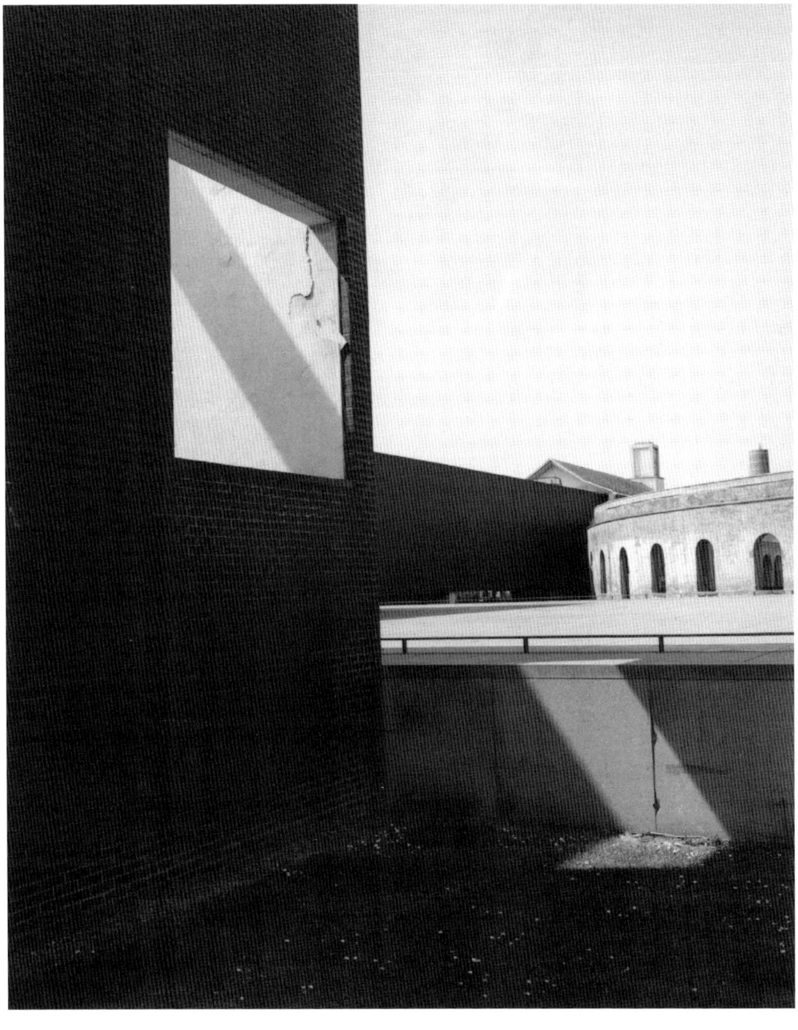

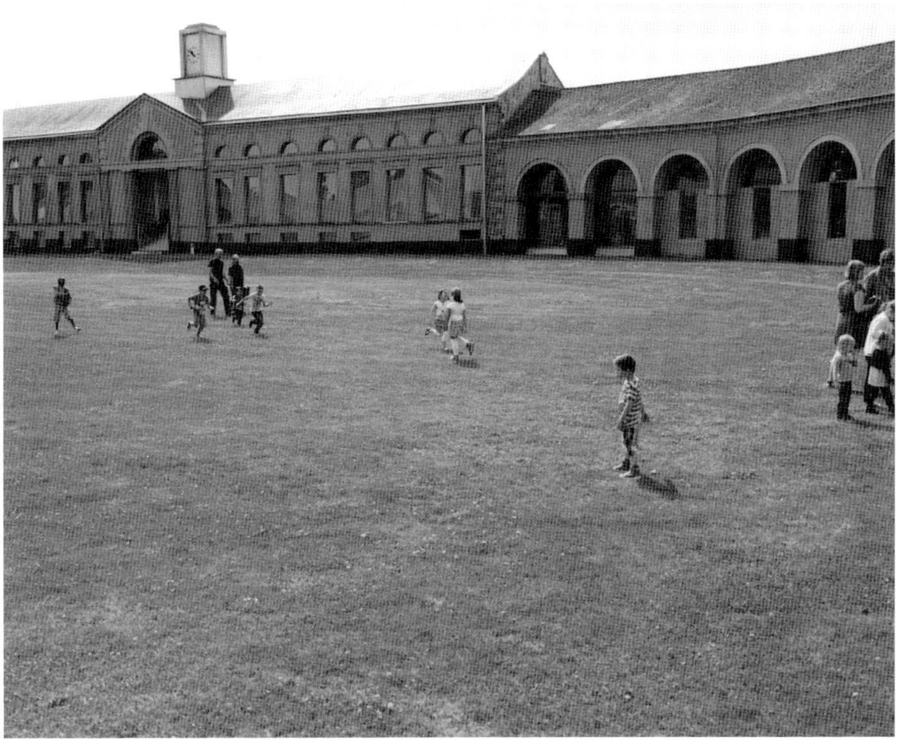

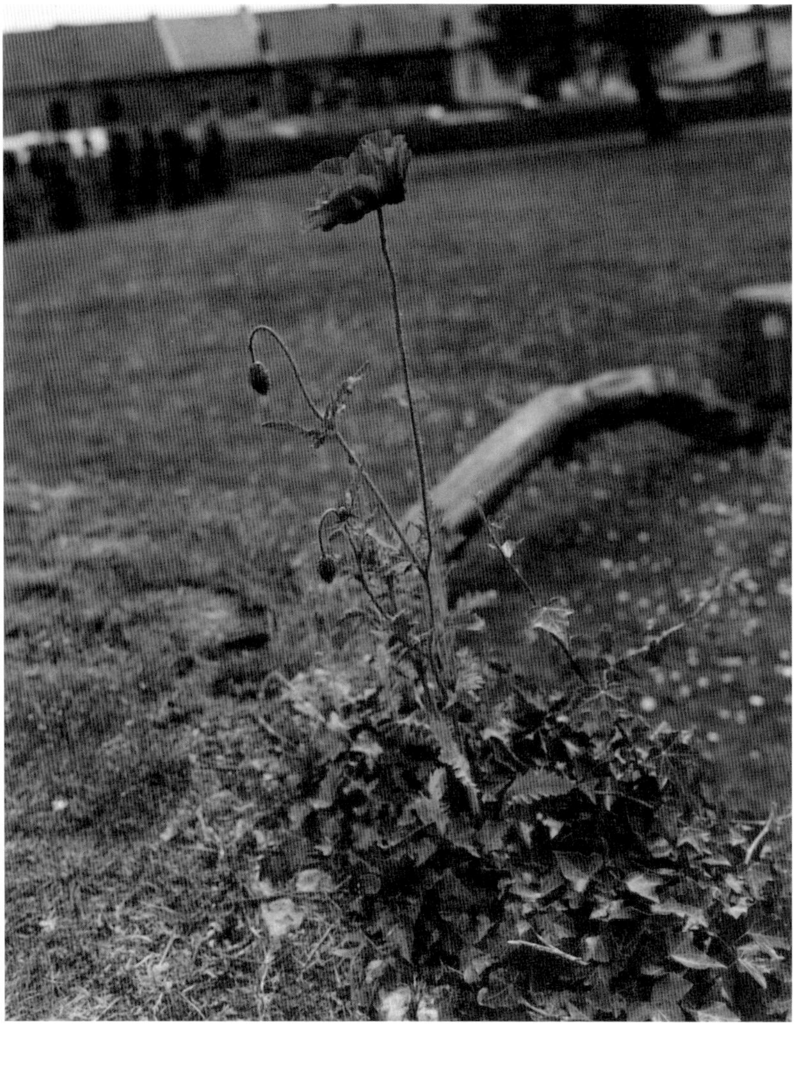

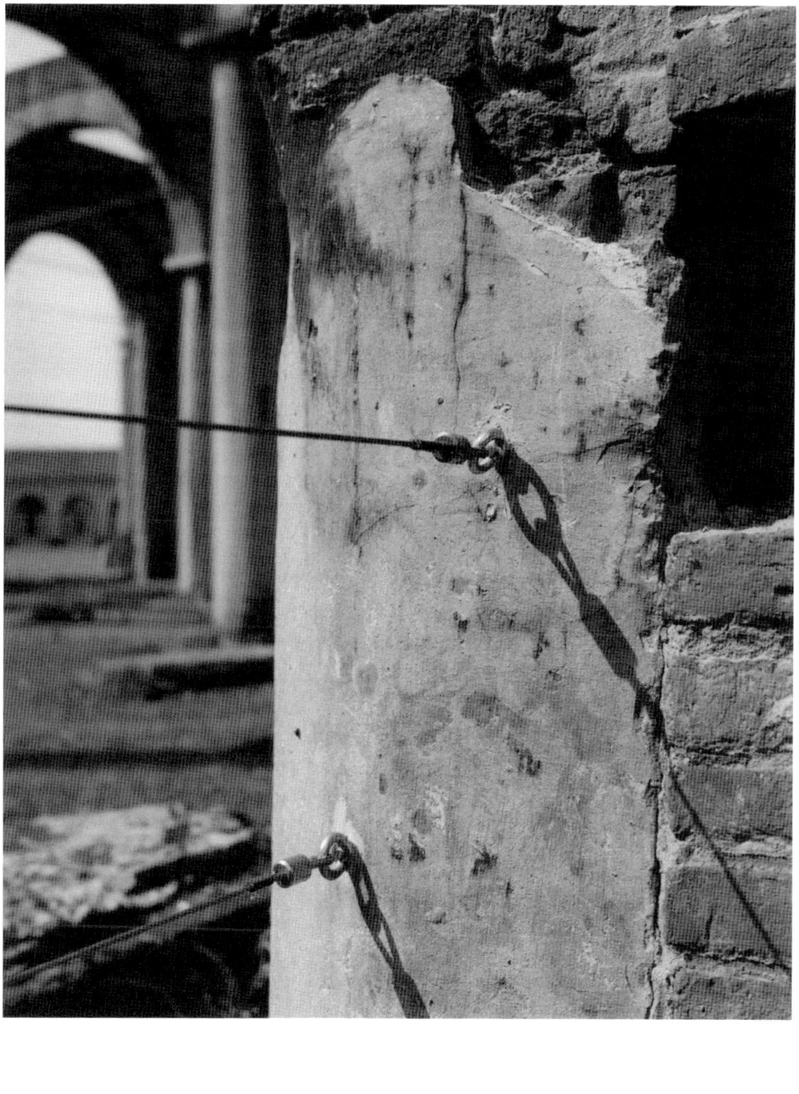

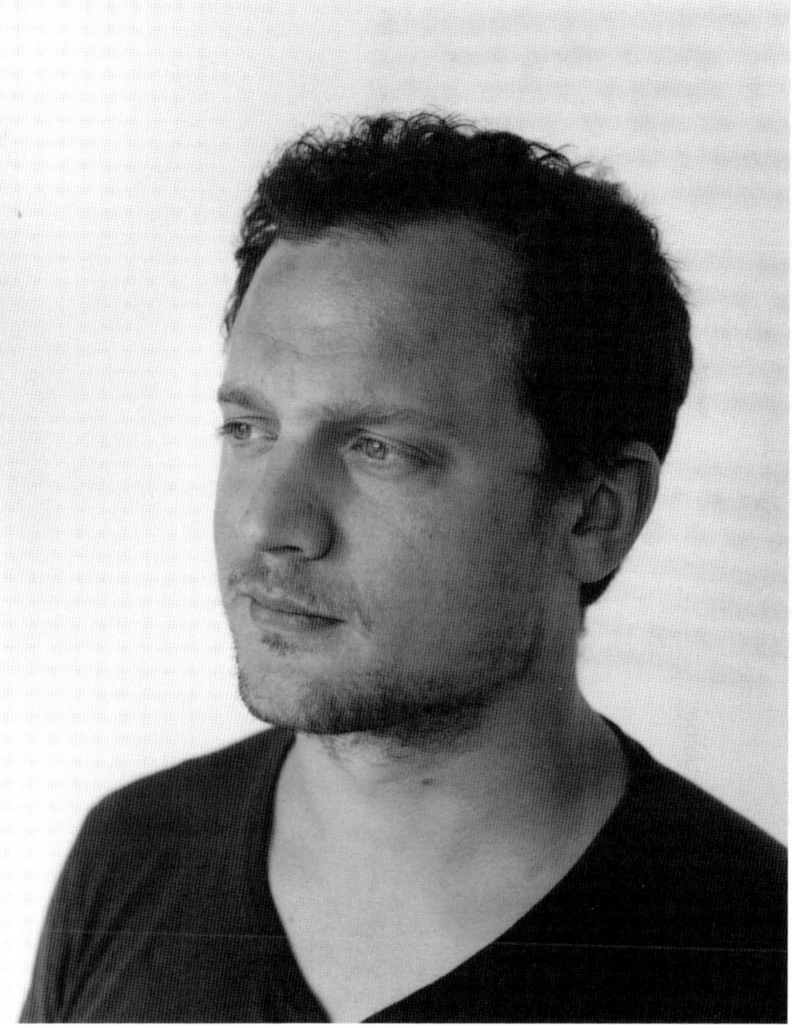

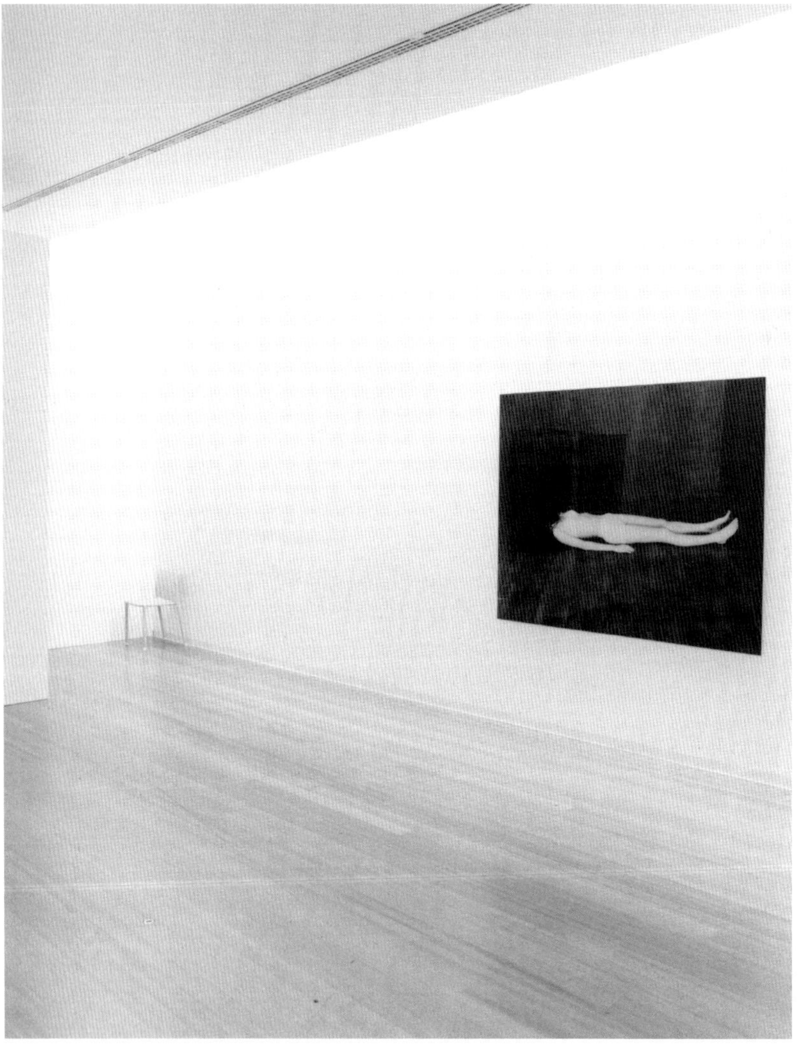

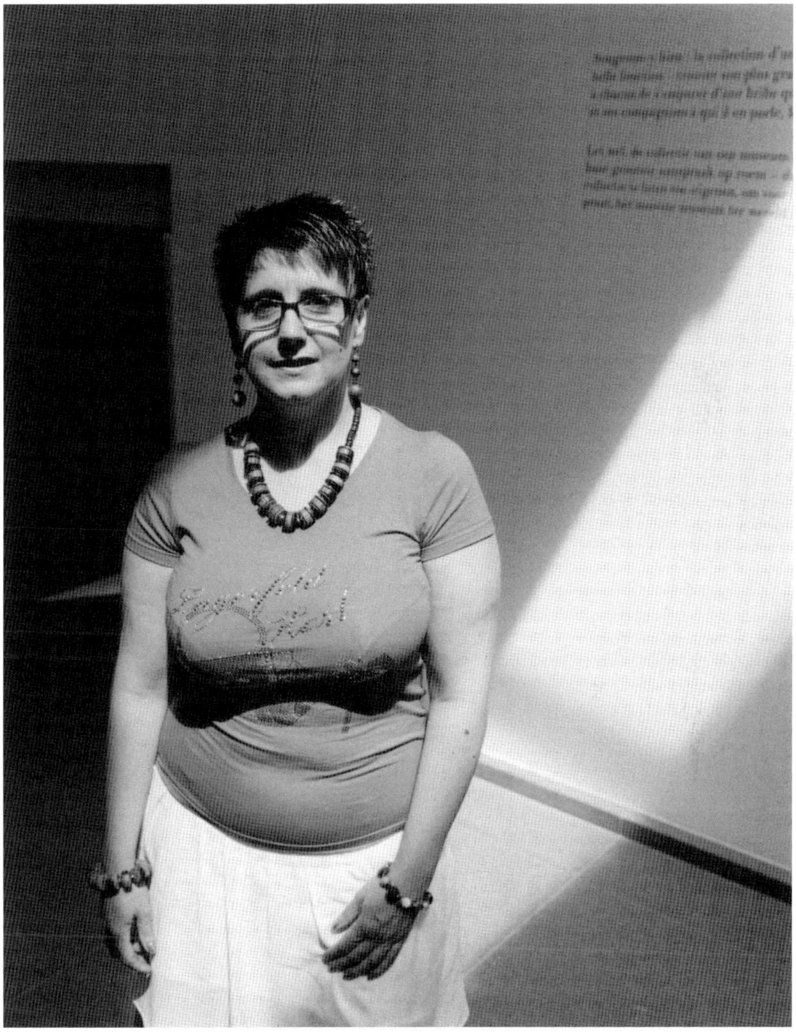

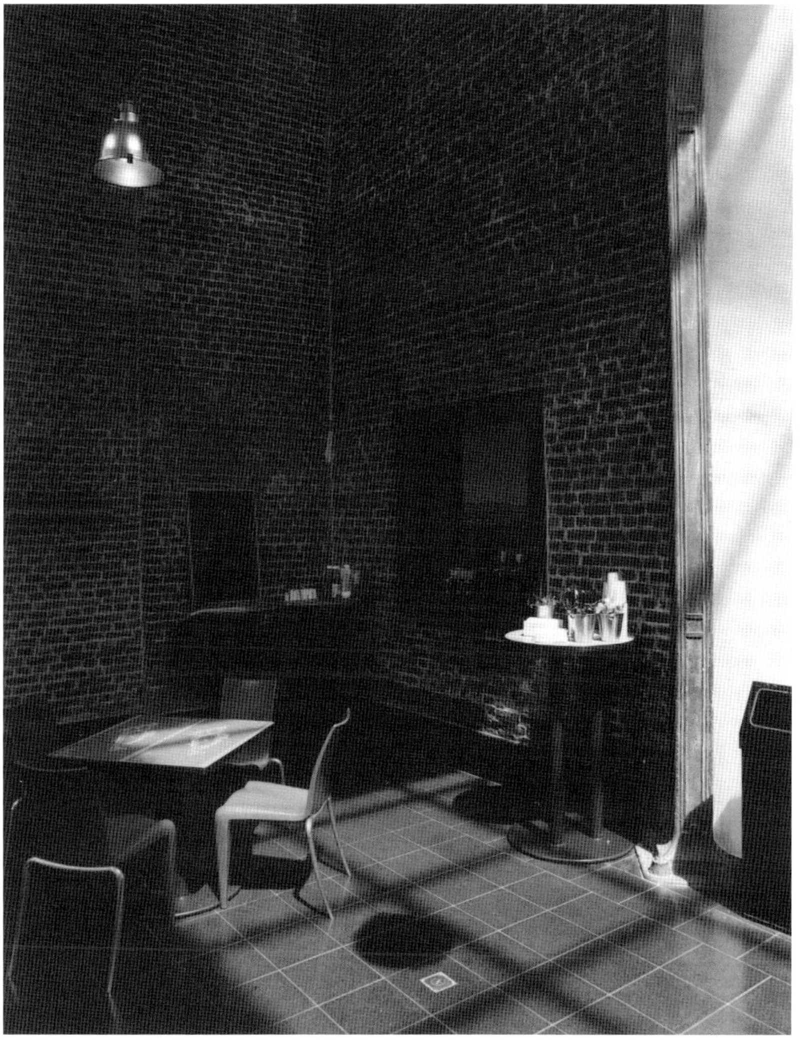

Monuments

This scene is a tour of two of the former mining sites now part of the UNESCO heritage declaration: the Fosse d'Arenberg, in Wallers and the Fosse 9–9bis, in Oignies. The photographs include interiors and details of machinery and mining tools.

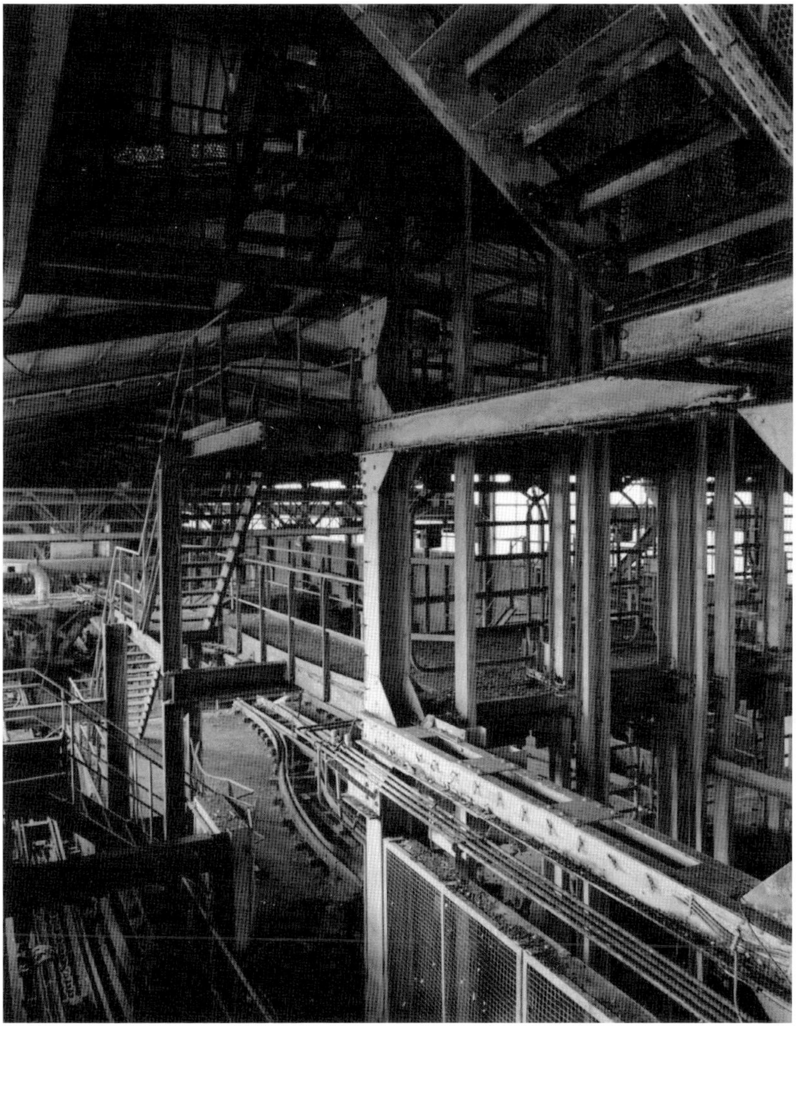

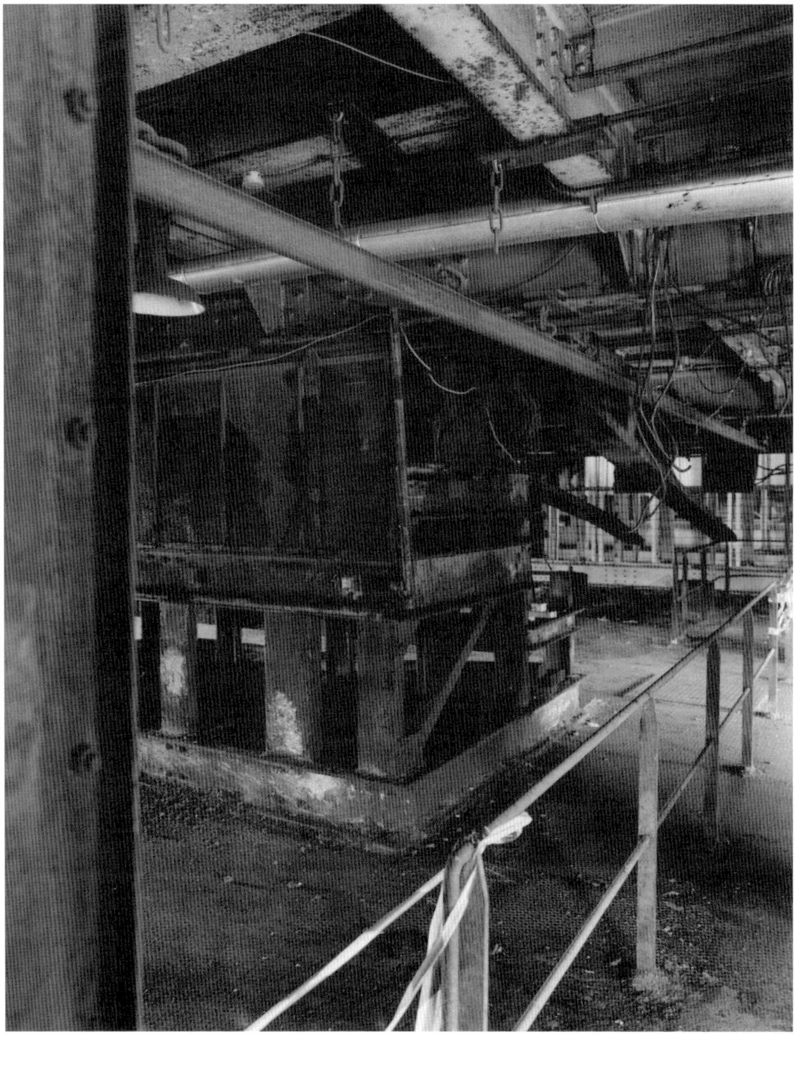

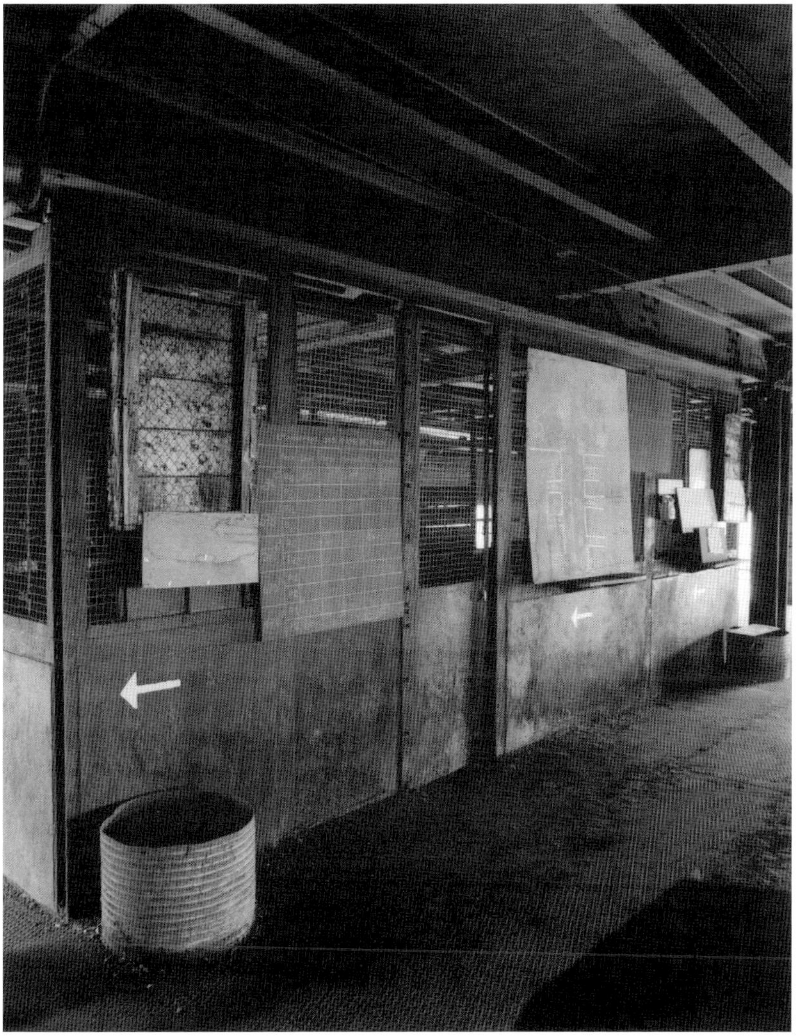

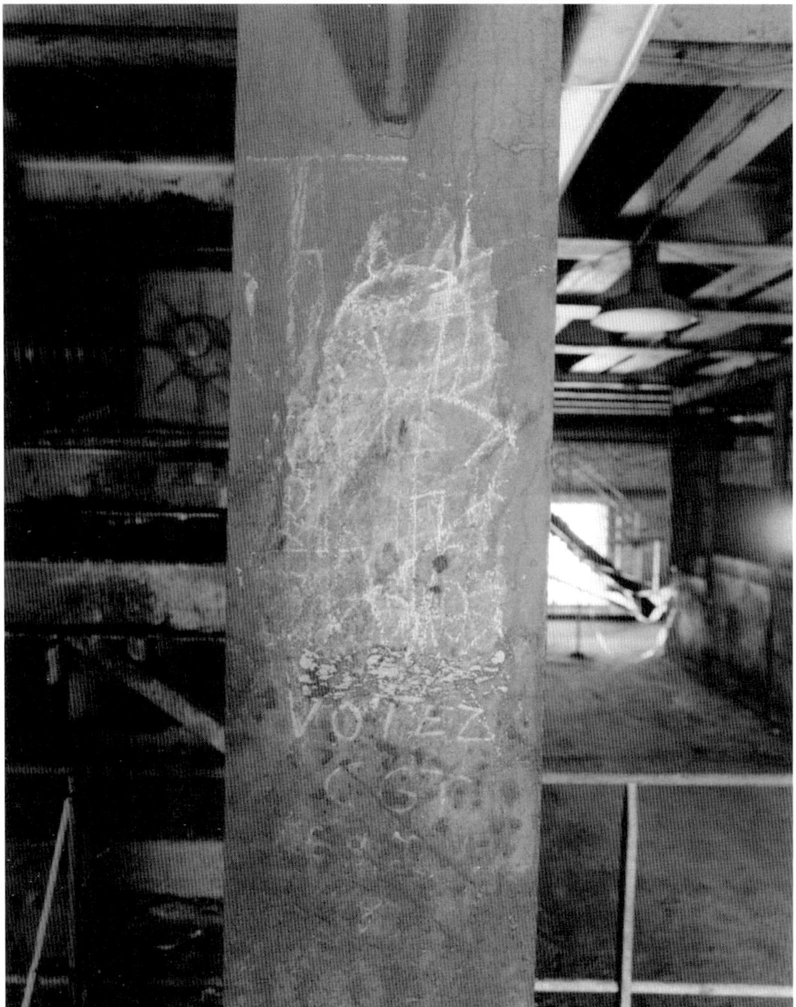
VOTEZ

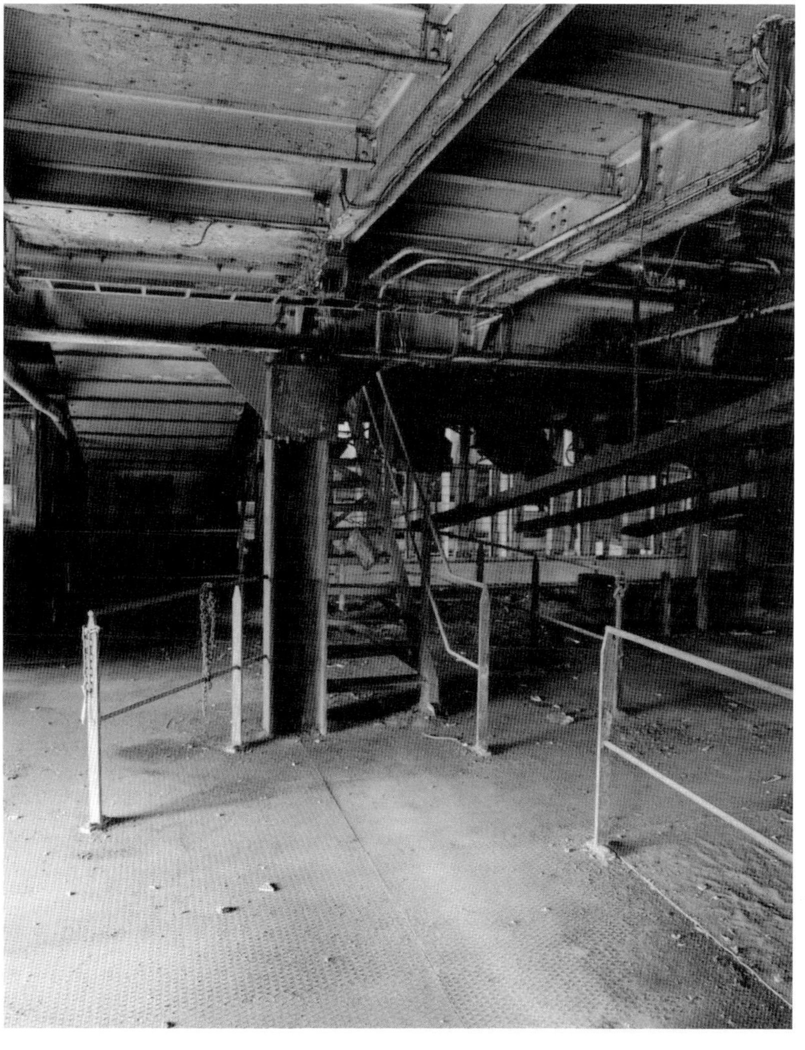

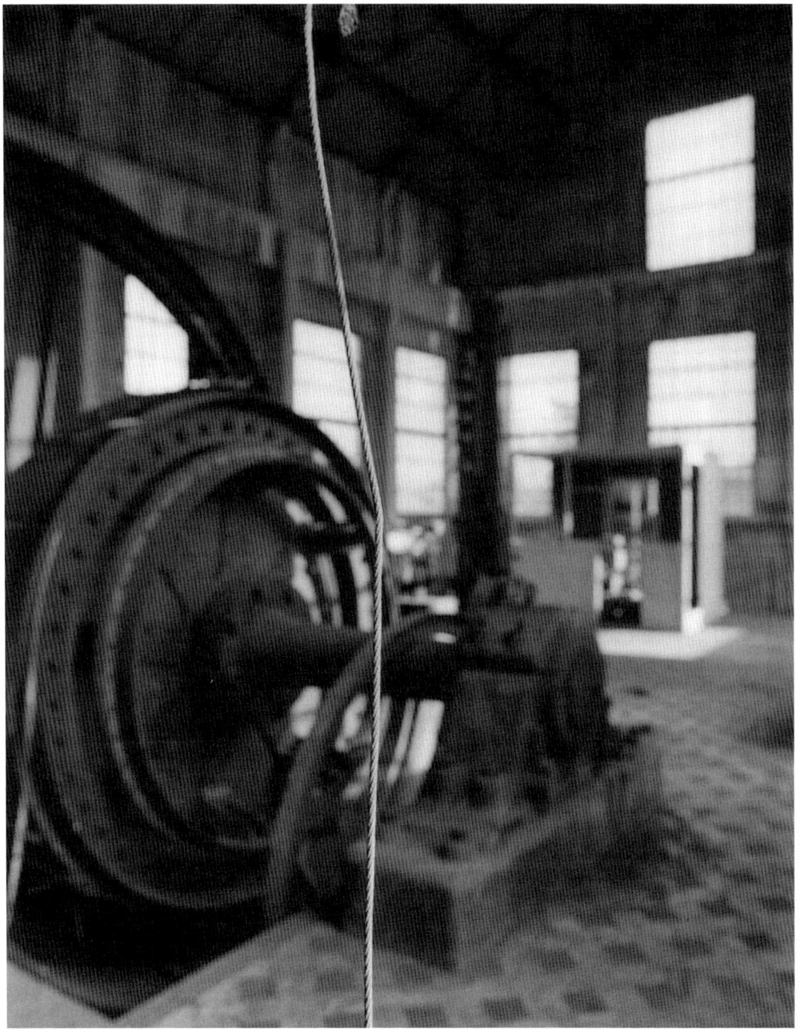

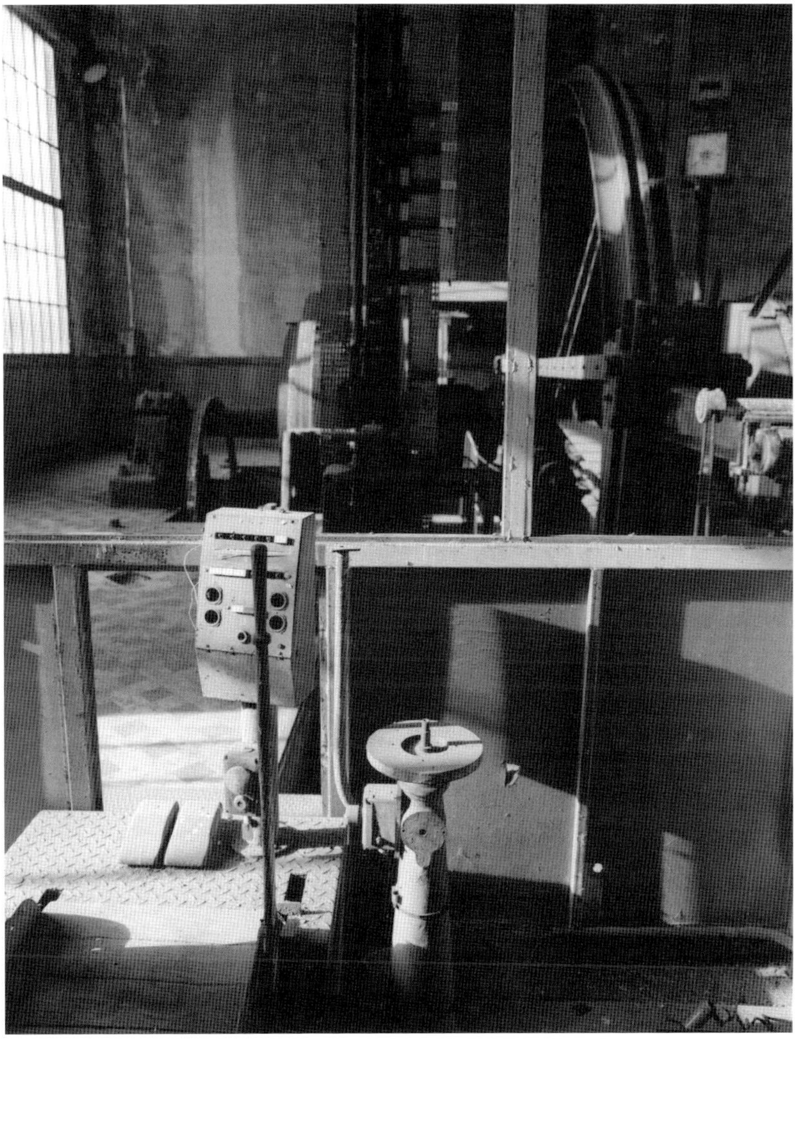

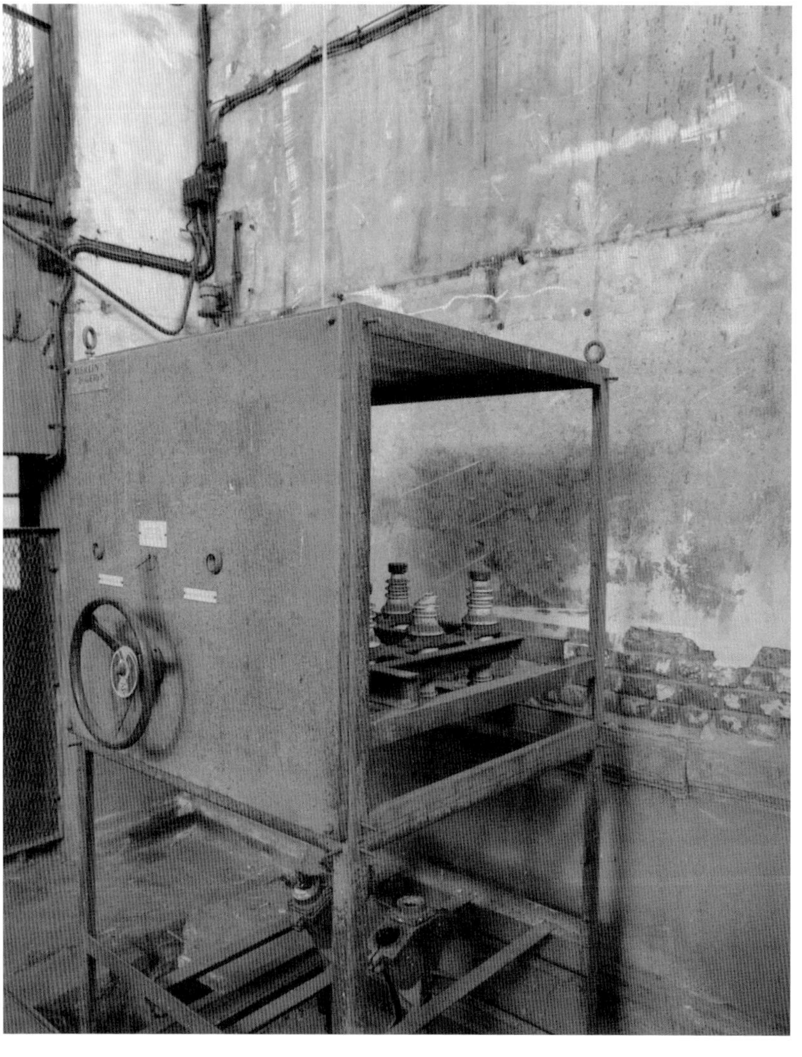

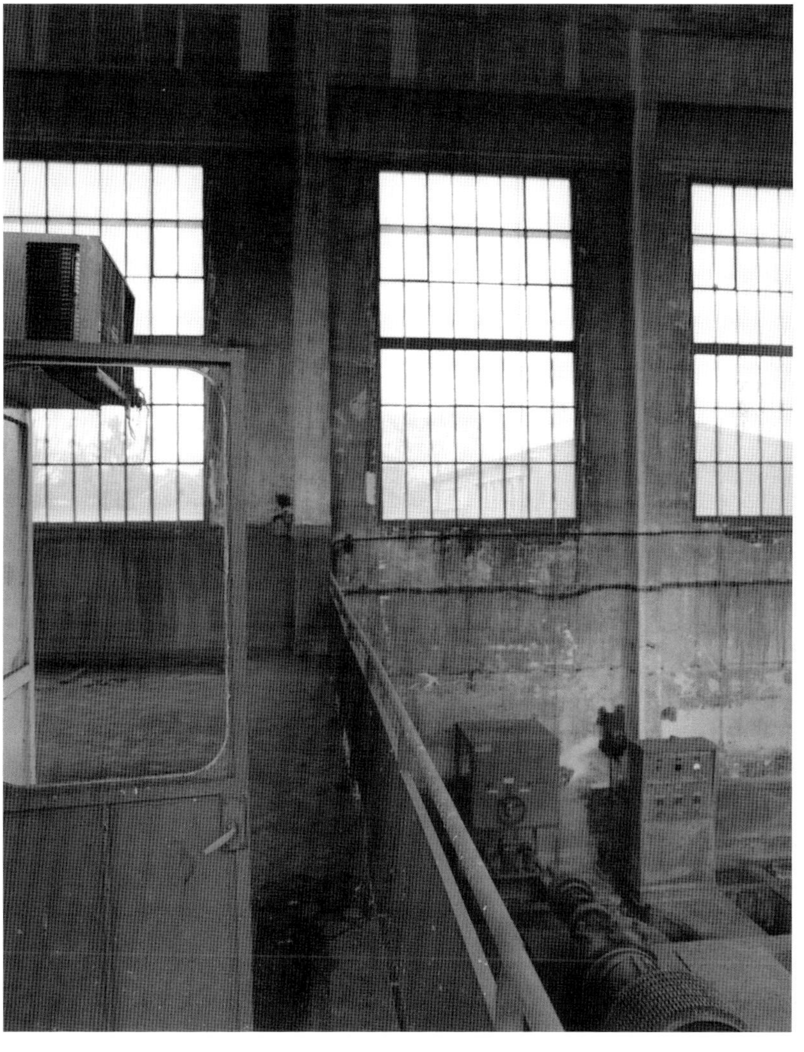

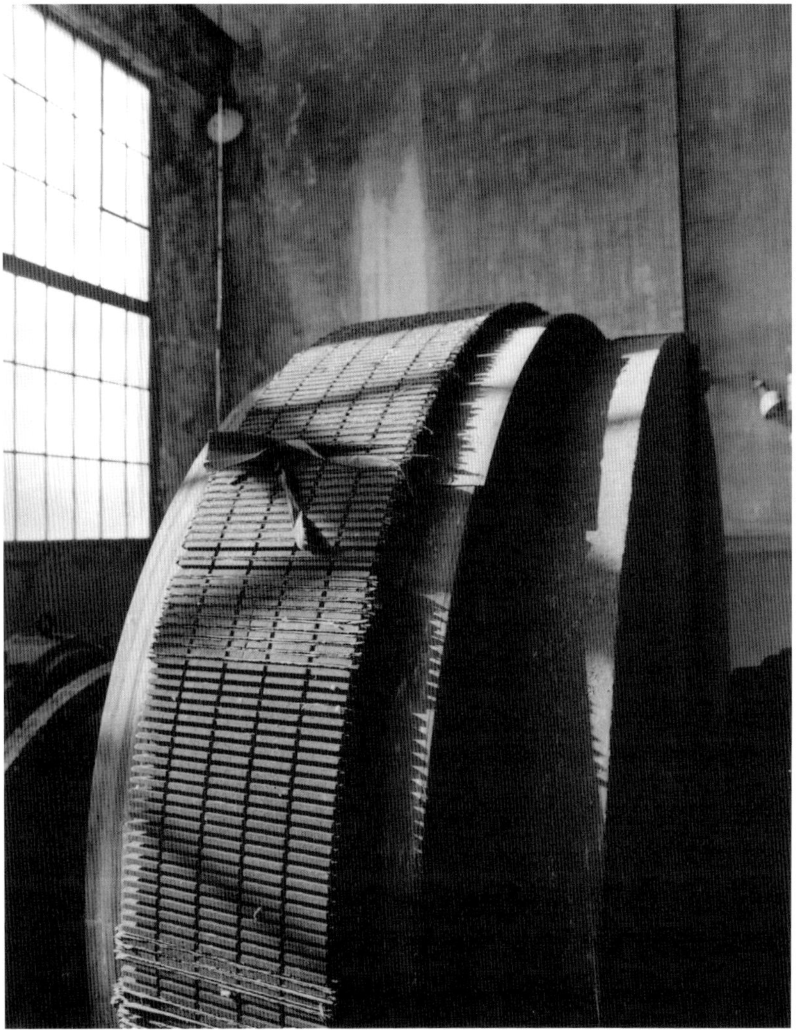

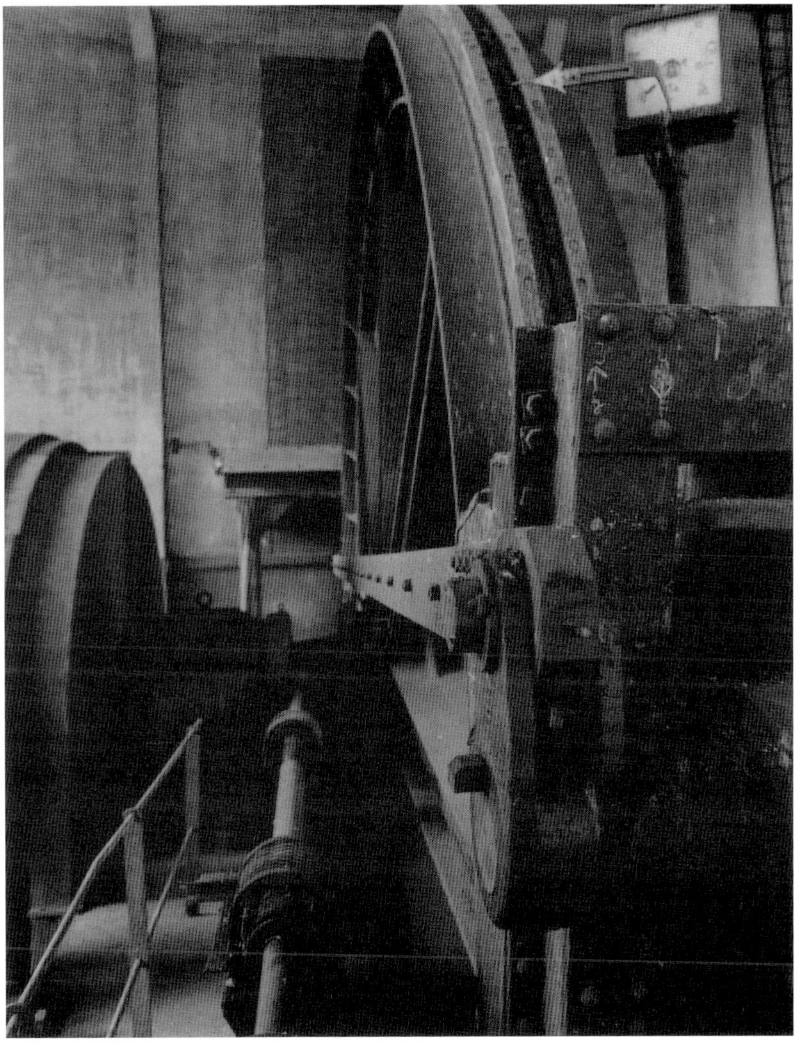

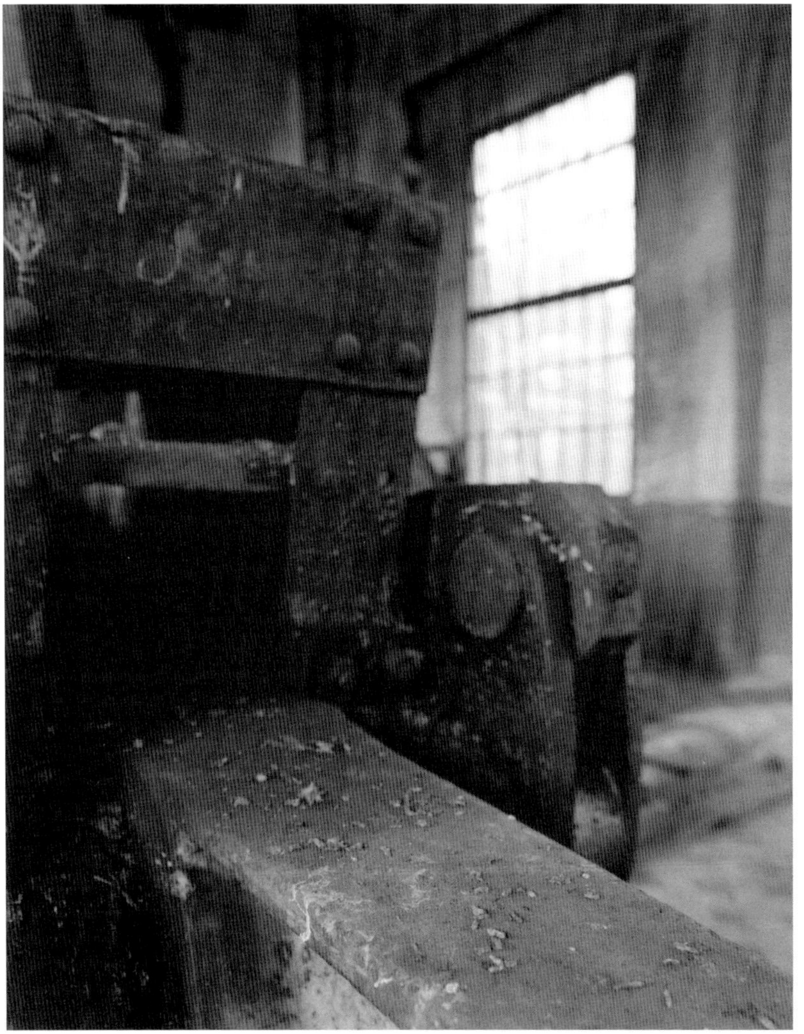

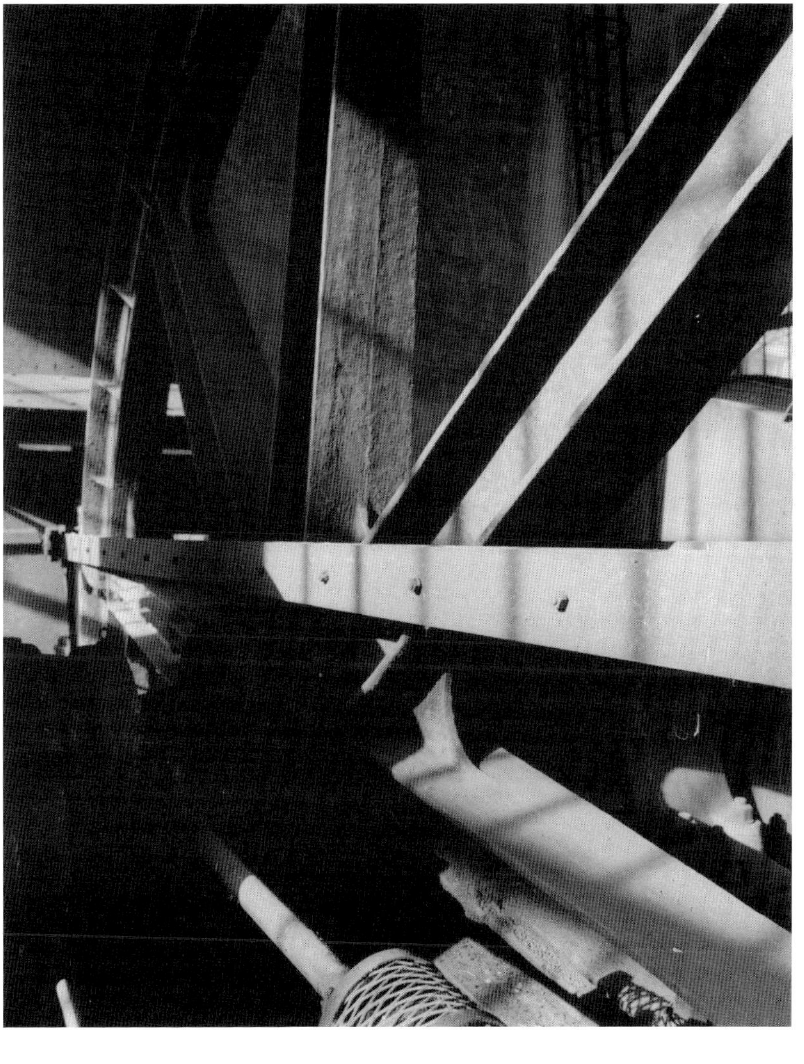

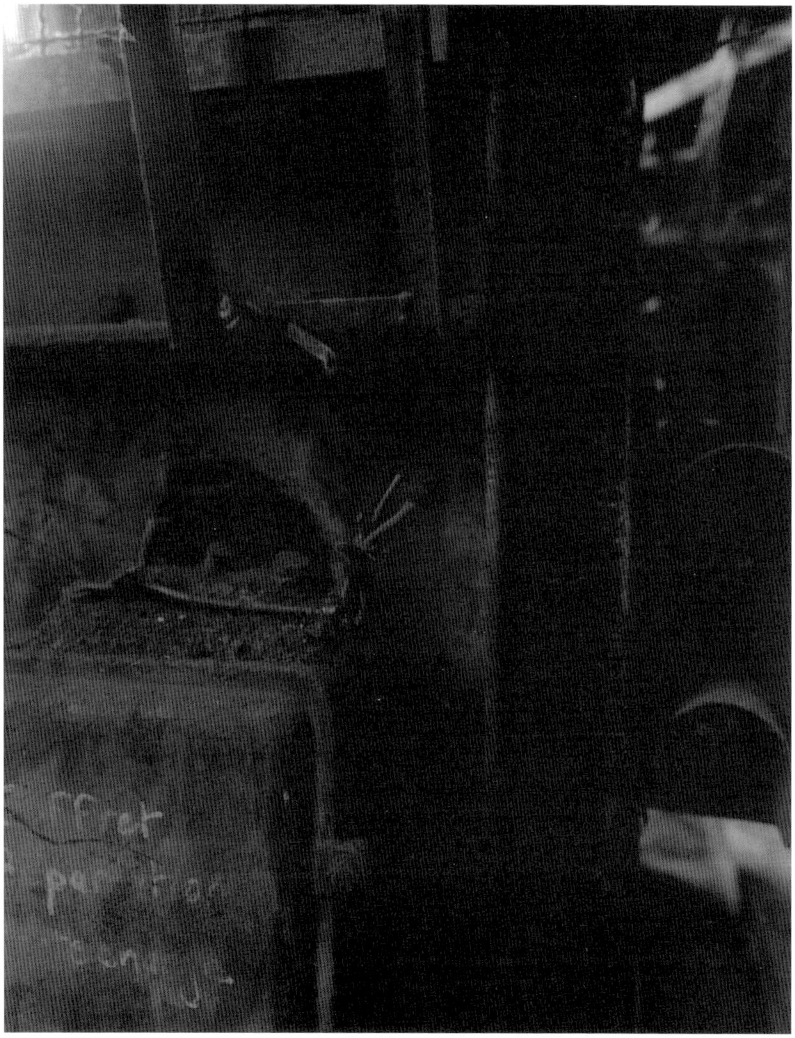

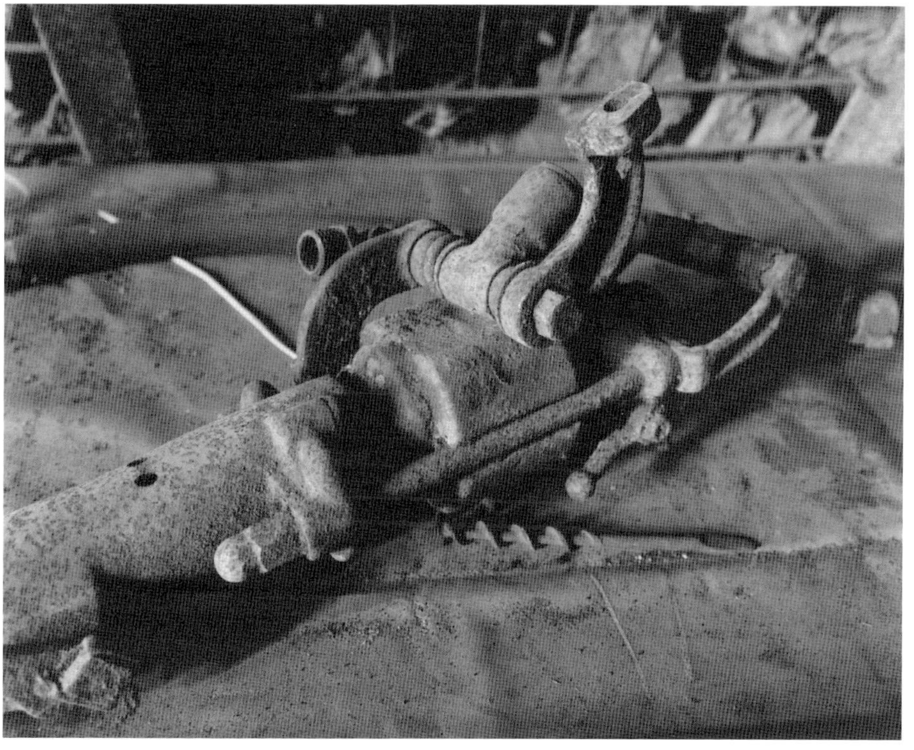

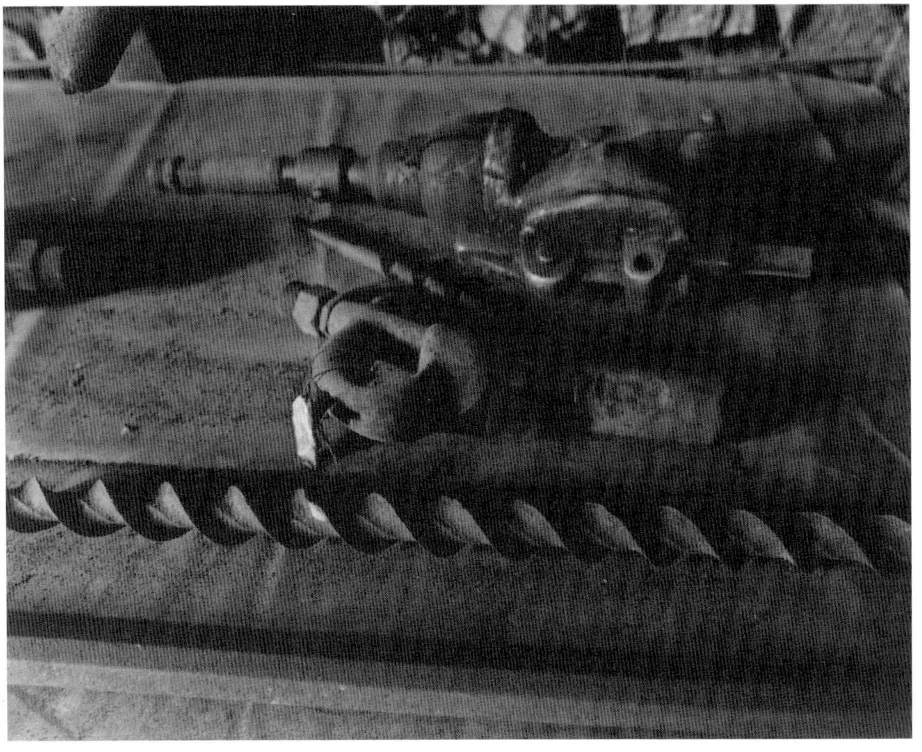

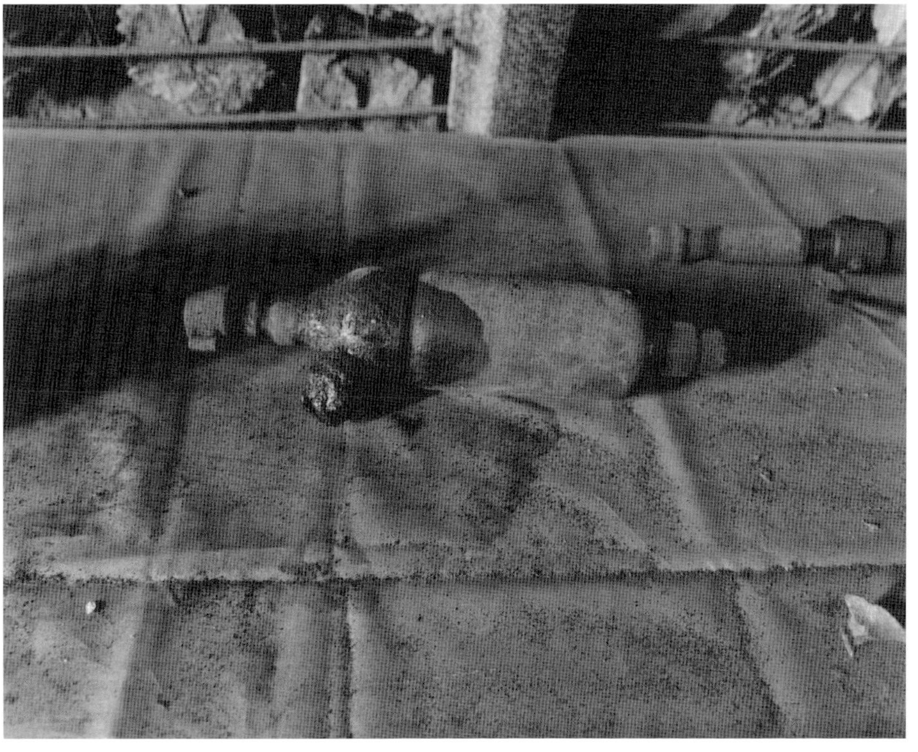

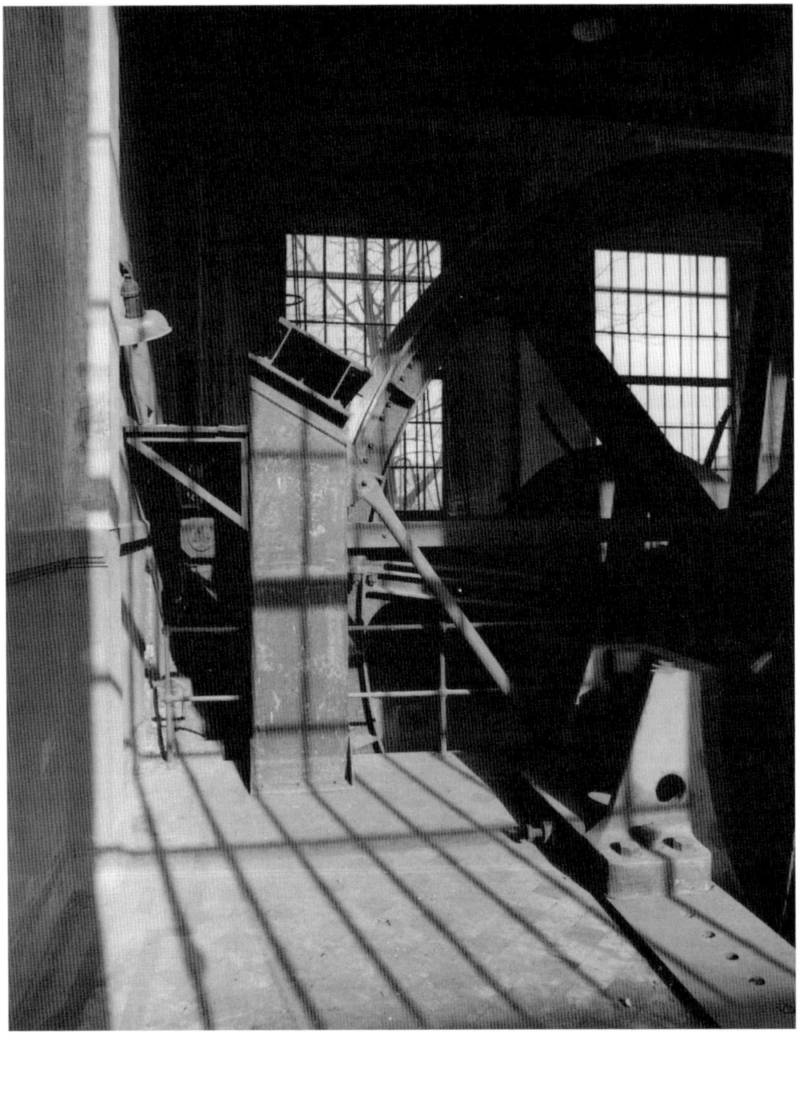

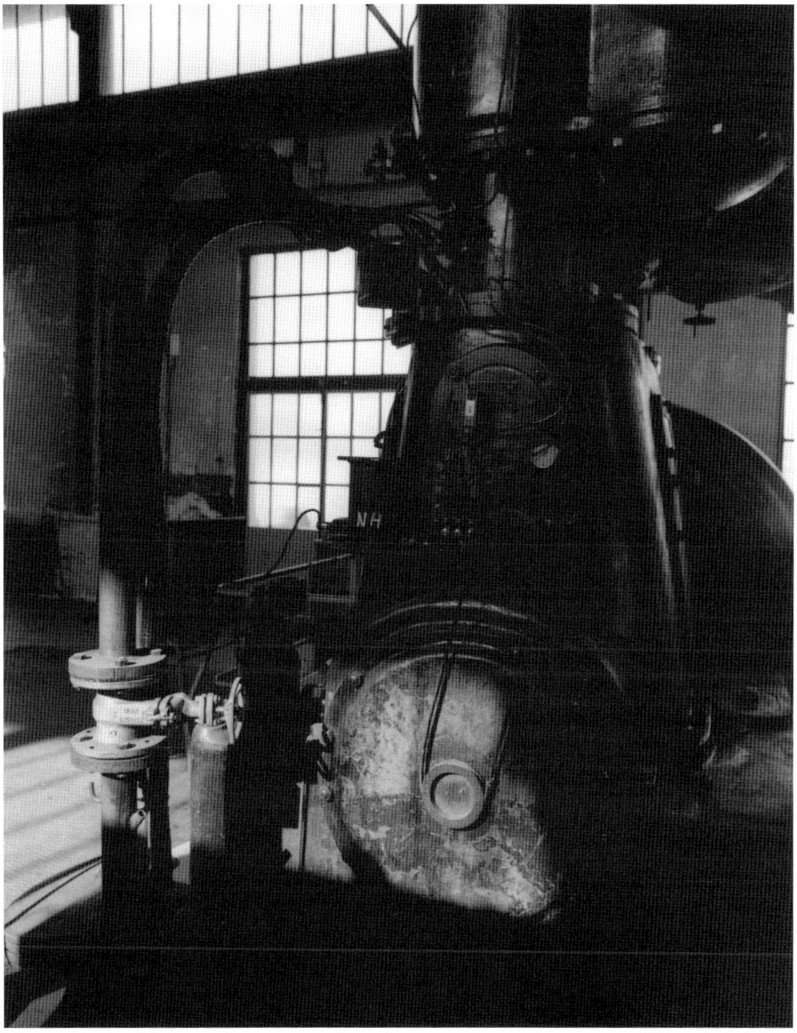
NH

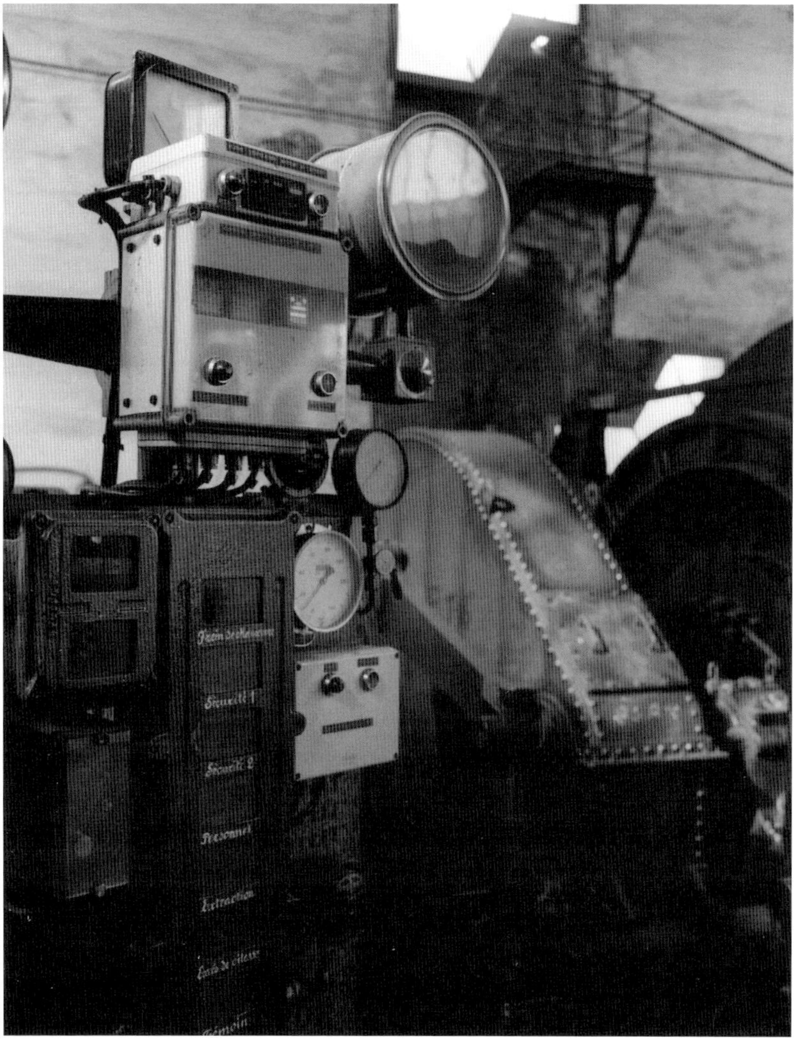
Personnel
Extraction

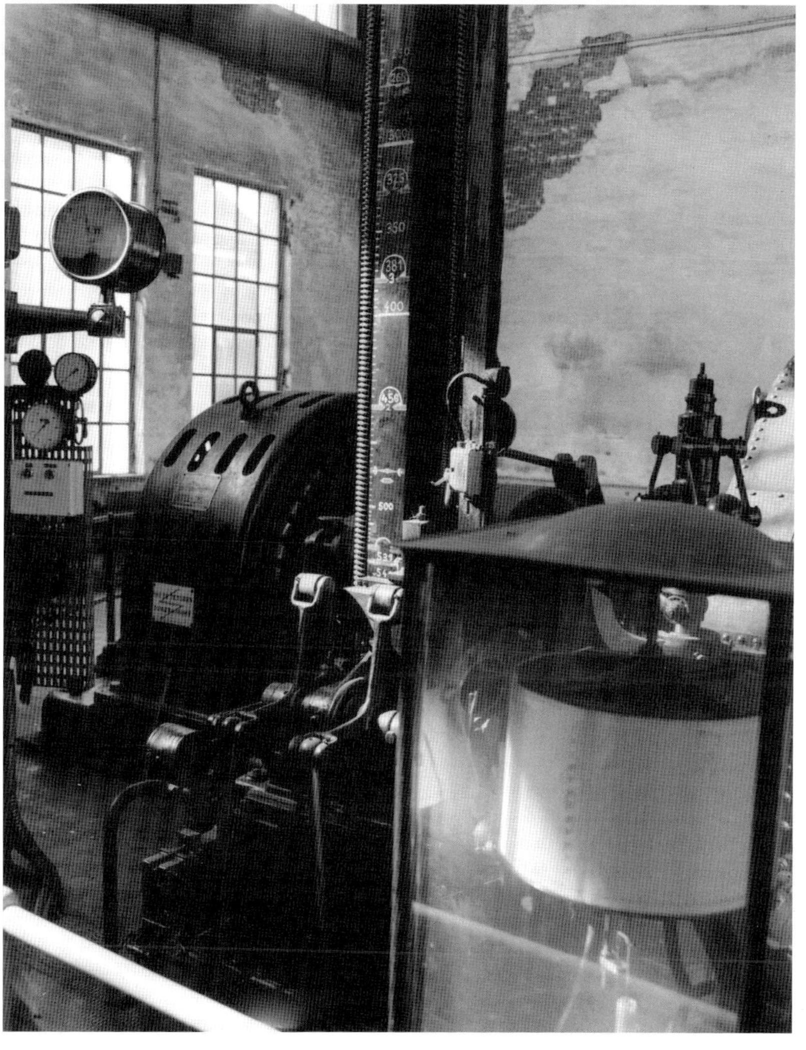
350
400
500

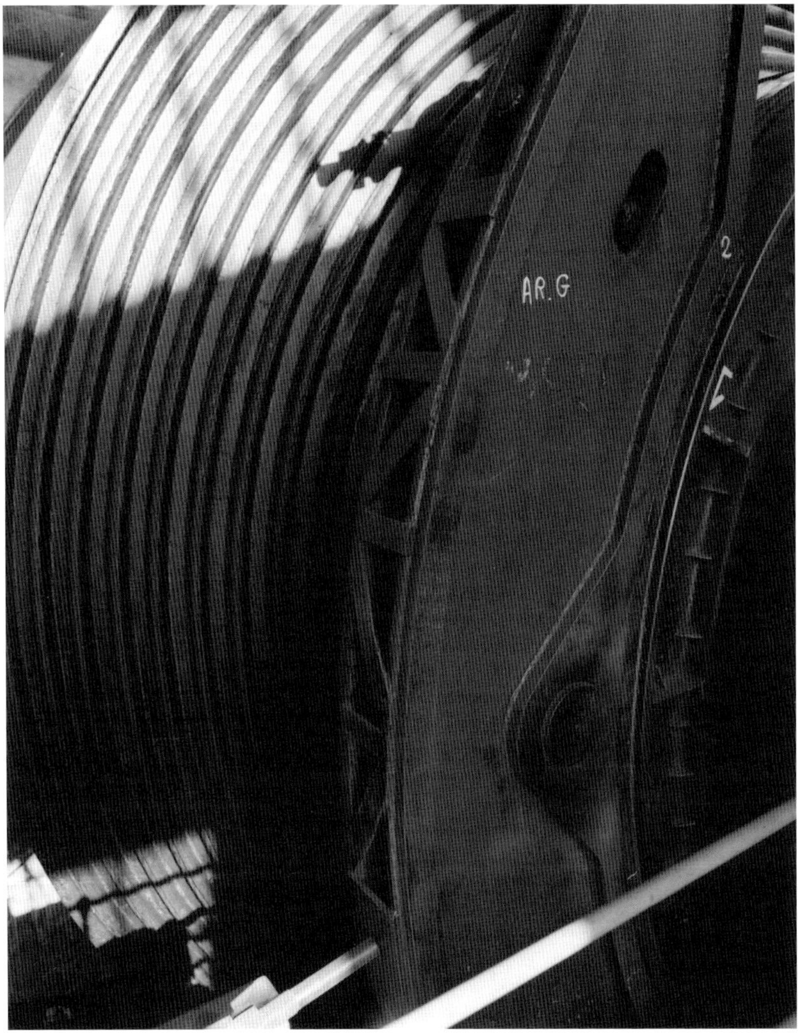
AR.G
2

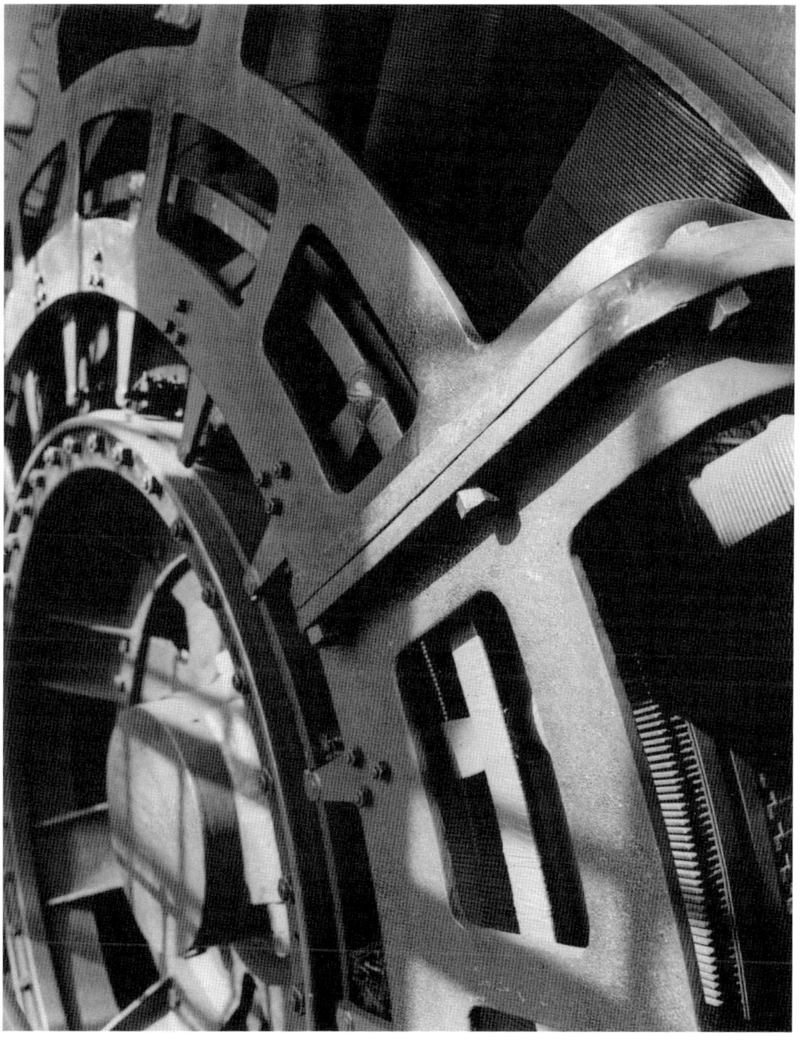

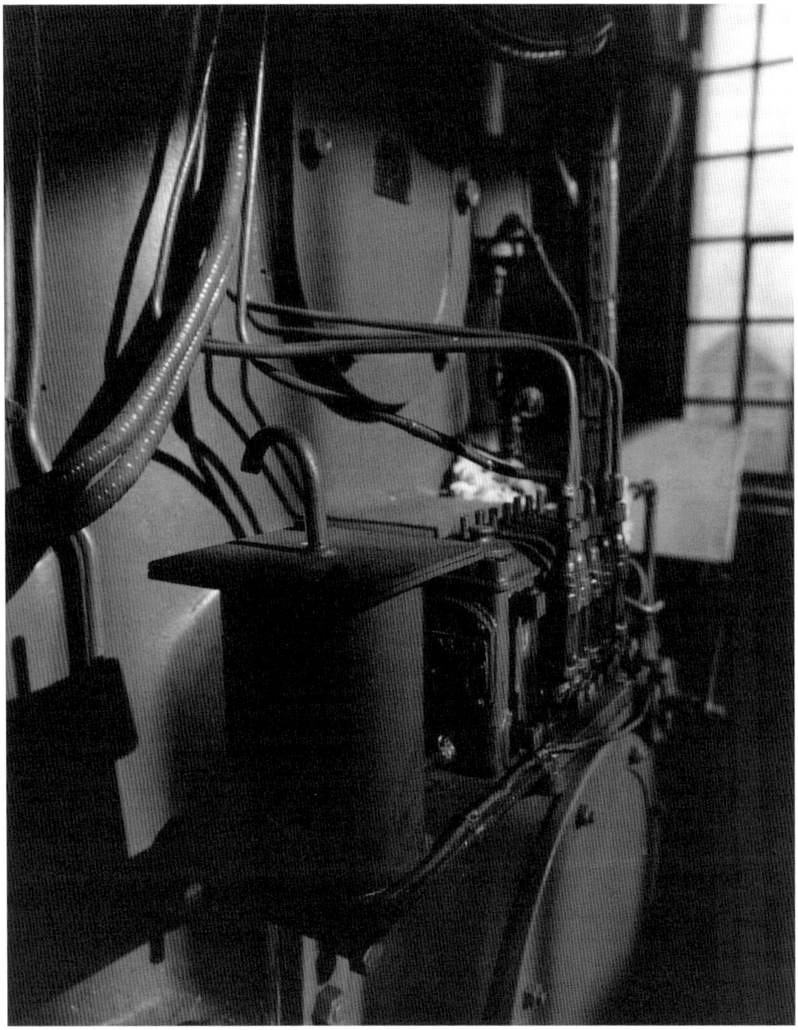

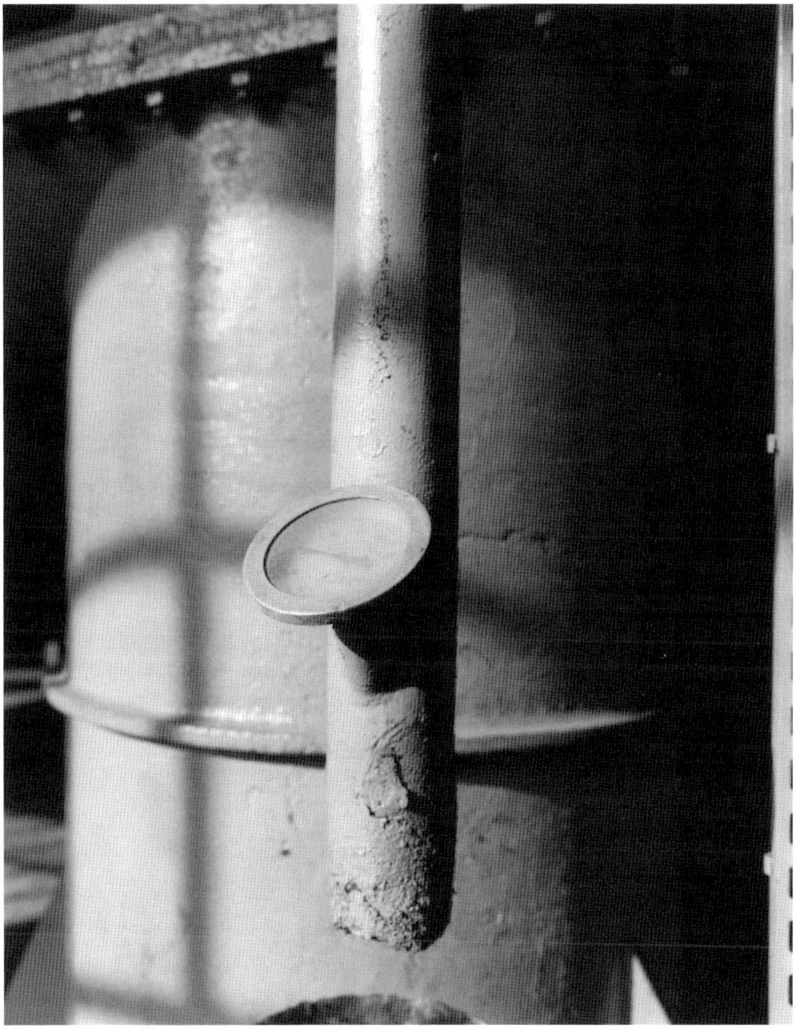

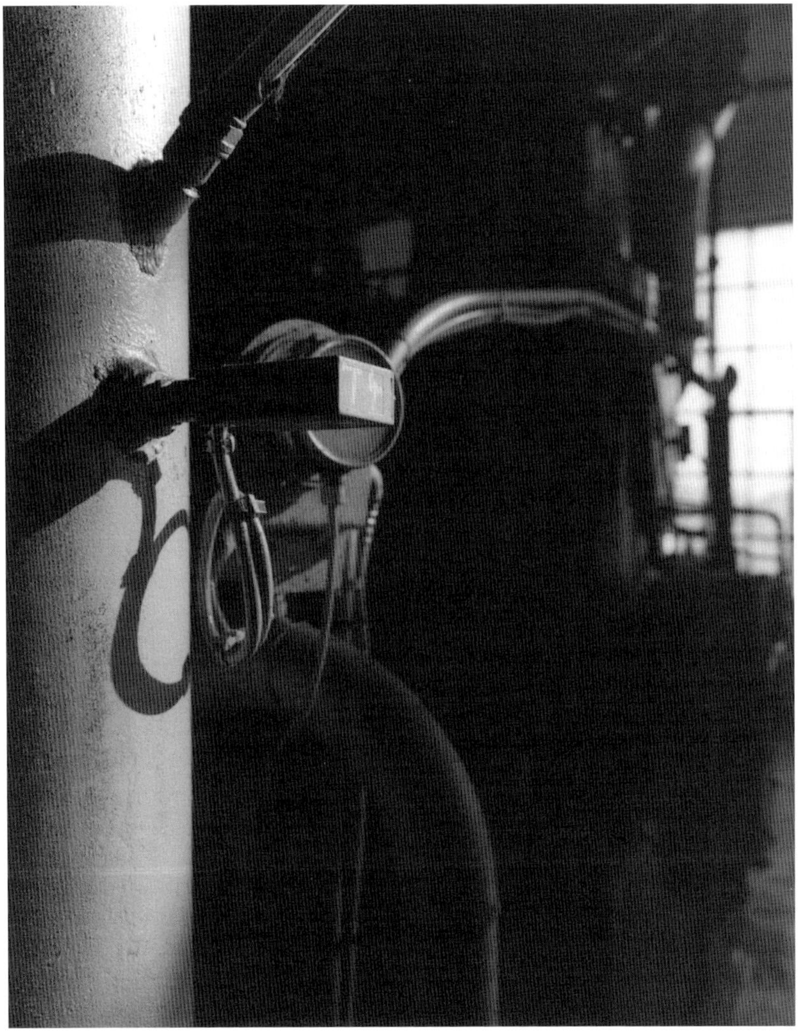

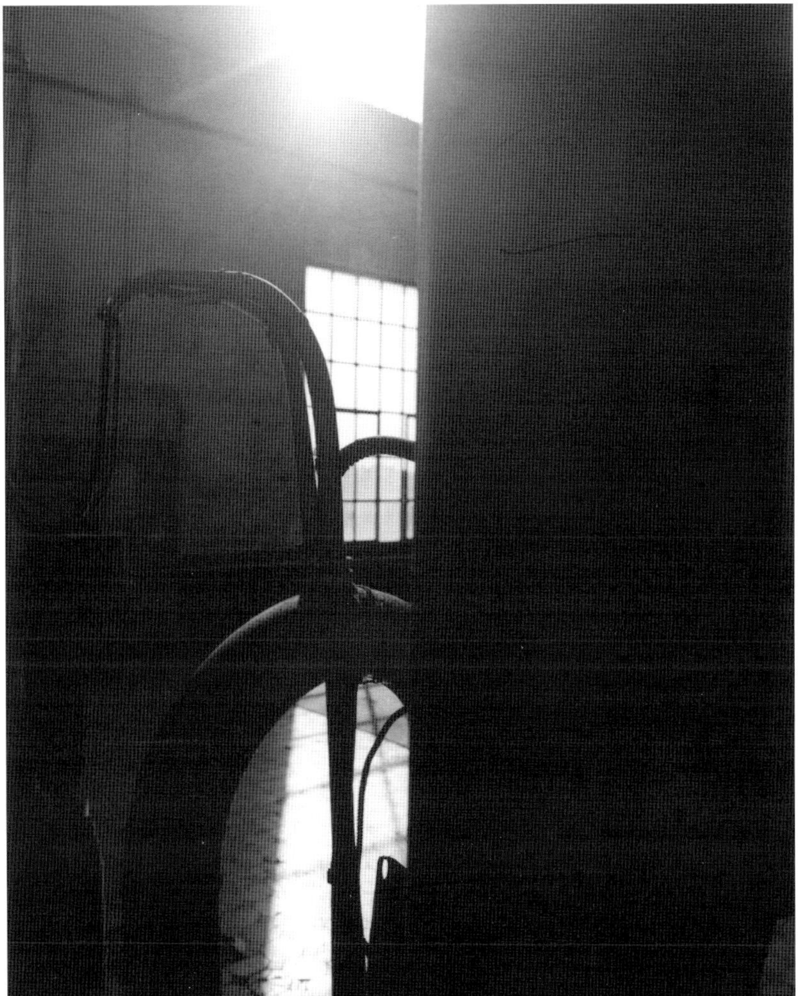

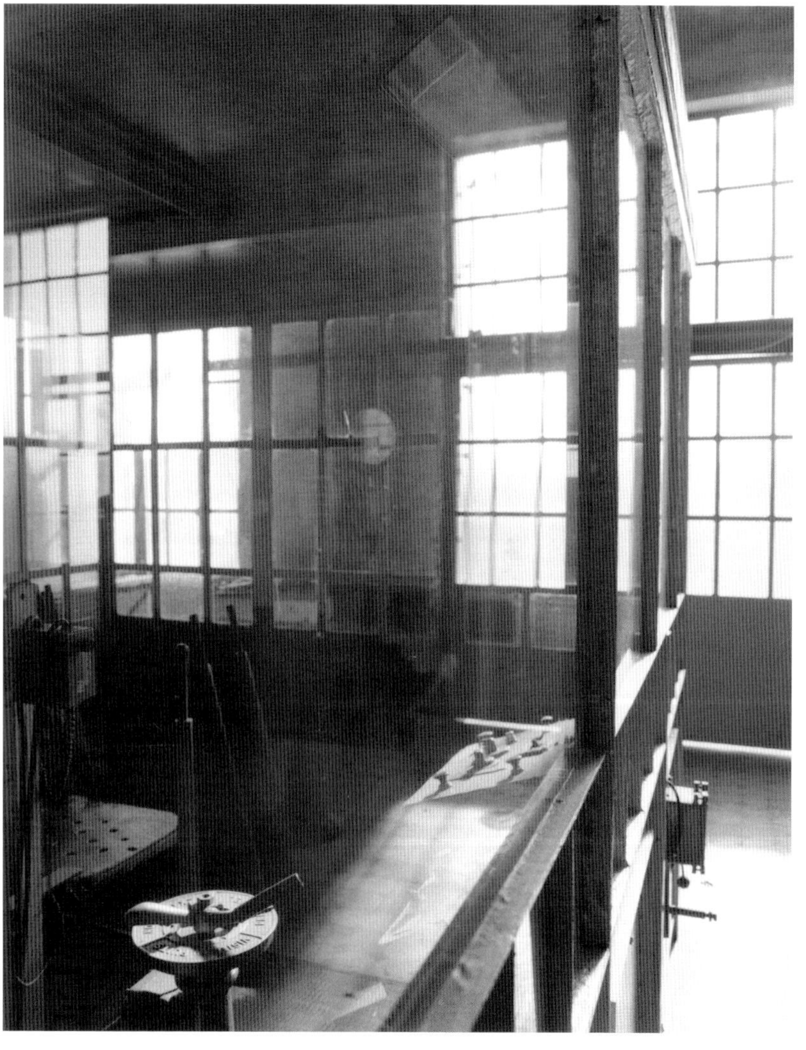

Mining Museum, Lewarde

The Centre Historique Minier de Lewarde is the most important historic institution concerning coal industry in the Nord-Pas-de-Calais region. It was founded in 1982 and opened three years later, in the former mining site of the fosse Delloye. Its originary function was to be the storage of the official archives of the mining companies of the coalfield before and after the nationalization in 1946. The mining activity stopped here in 1971 and the galleries were closed. The museum is funded by the Ministry of Culture and the Nord-Pas-de-Calais regional administrations, and managed by the Centre Historique Minier de Lewarde association. The Centre holds a collection, archive and library and keeps an important research, education and exhibition activity. The permanent exhibition includes some dioramas and presents a didactic history of coal and its uses, a history of coal mining and of coal exploitation in the Nord-Pas-de-Calais region, a history of the site and the evolution of mining technologies since the 17th century until the 20th century. The visit includes the original mining facilities of the site and it ends with a dramatic descent to one of the galleries, which is actually a fake reconstruction that the visitor only discovers at the end.

The scene includes various details of the display and objects in the collection exhibition. It includes portraits of some museum workers, particularly of the guides to the visit, like Daniel Francke, former miner who is currently one of the most popular guides, also frequently represented in the publicity of the Centre. A younger visit guide is Anna Hembise, and Aurélie Bourgois is in charge of ticket sales and reception.

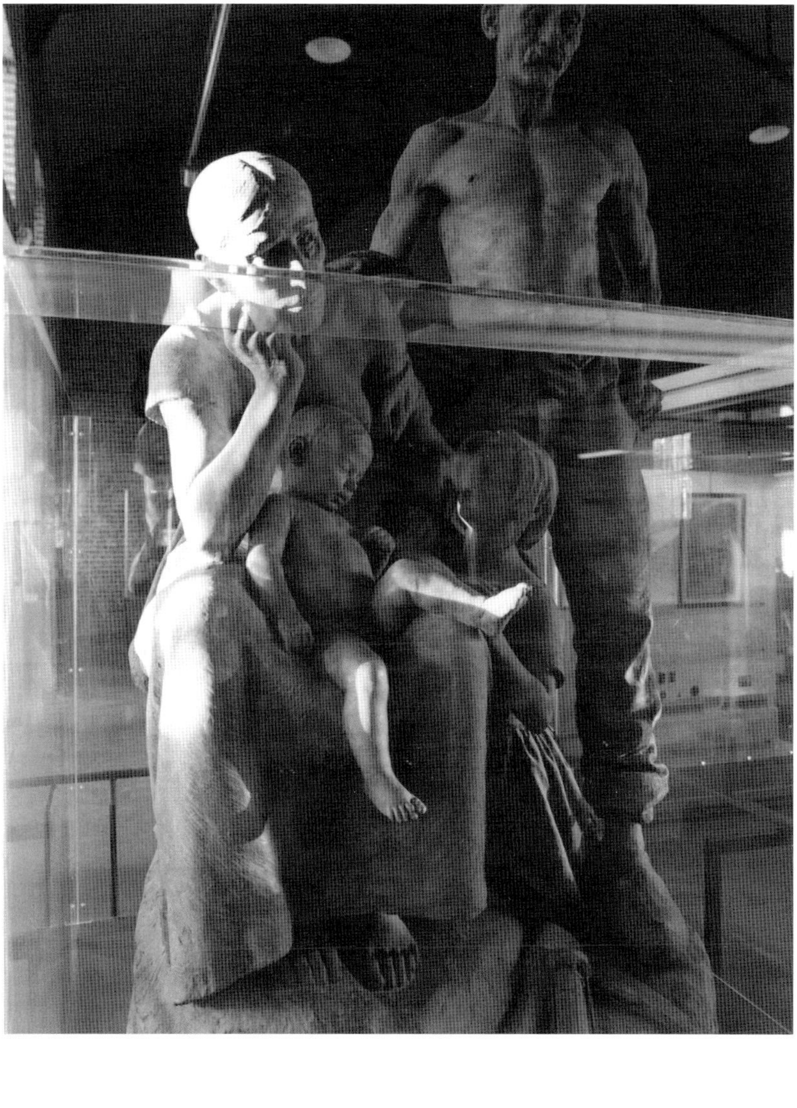

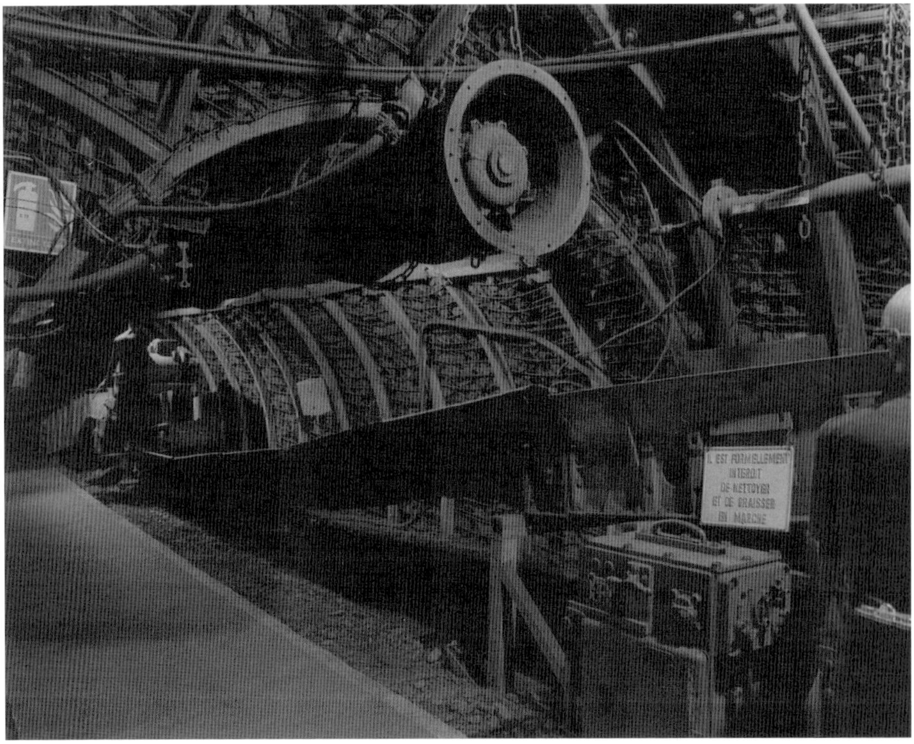
IL EST FORMELLEMENT
INTERDIT
DE NETTOYER
ET DE GRAISSER
EN MARCHE

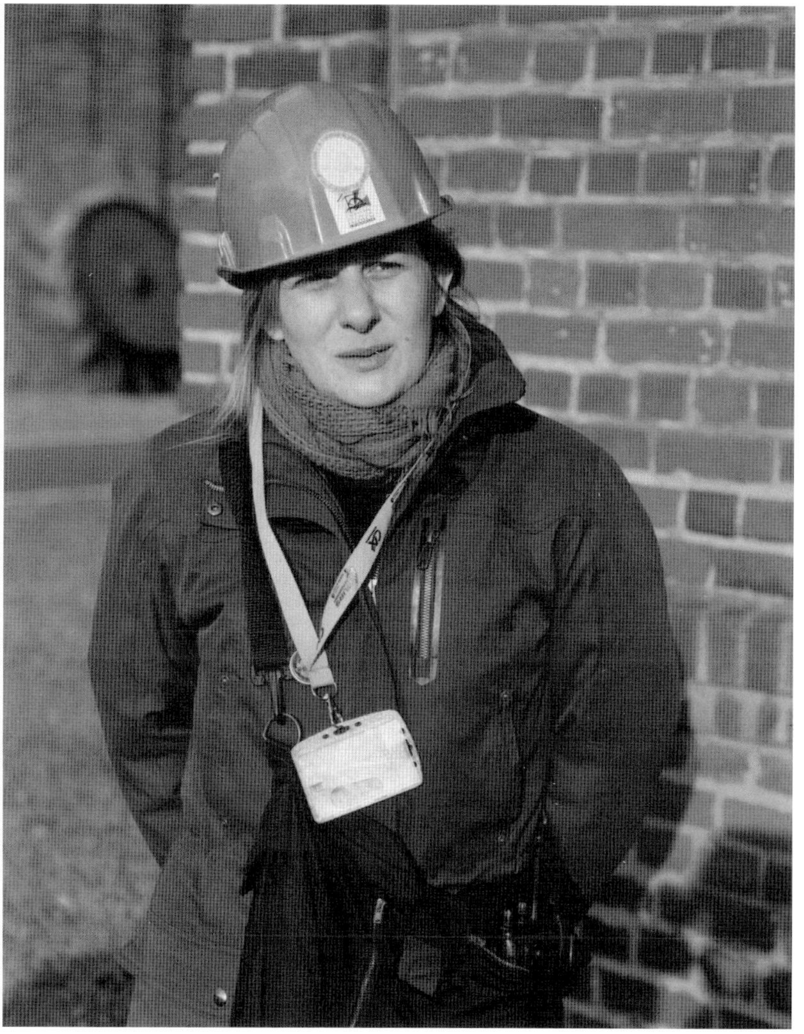

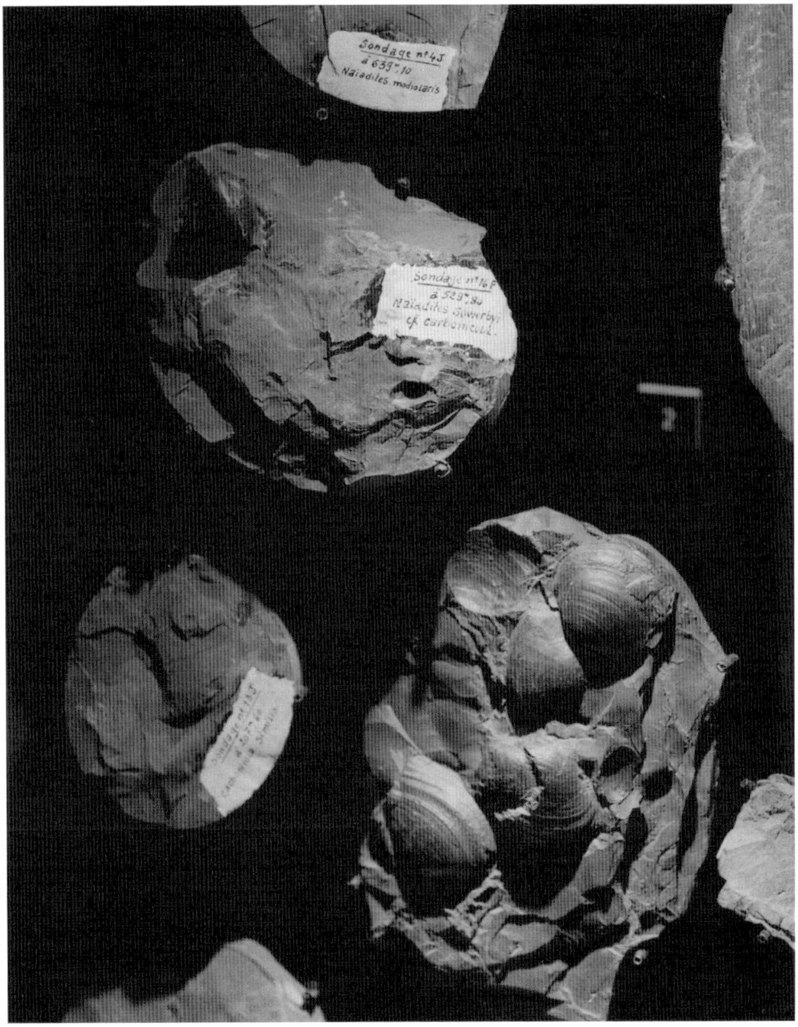
Sondage n°4J
à 639m,10
Naïadites modiolaris
Sondage n°16F
à 529m,80

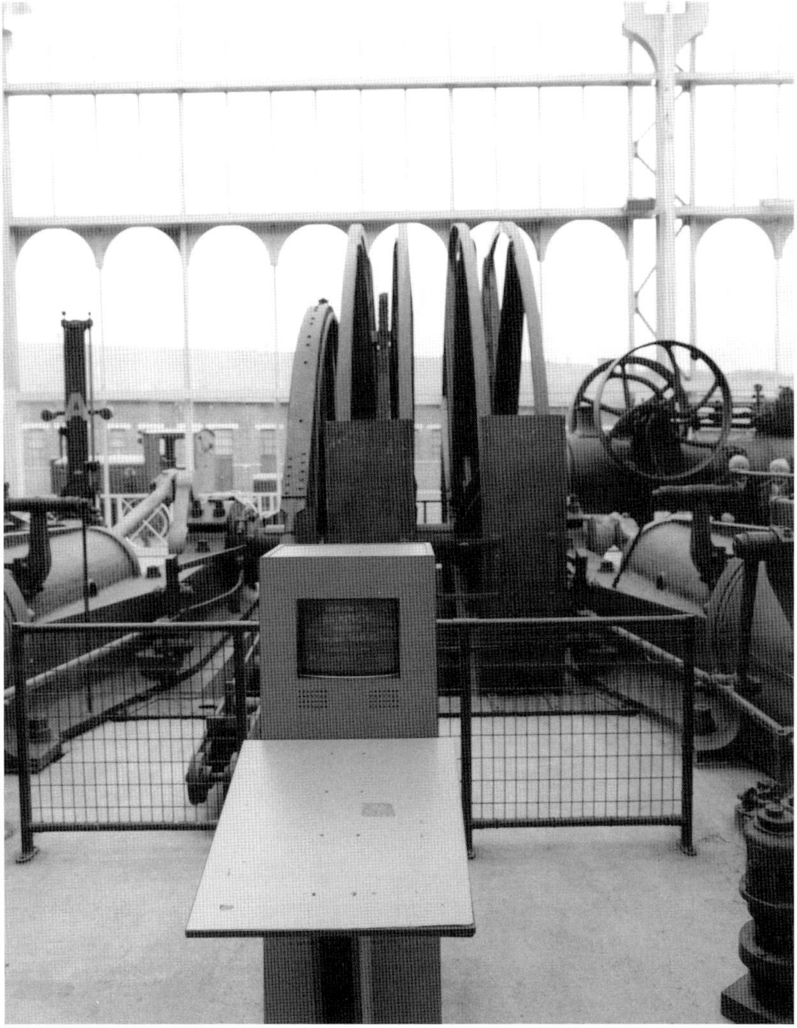

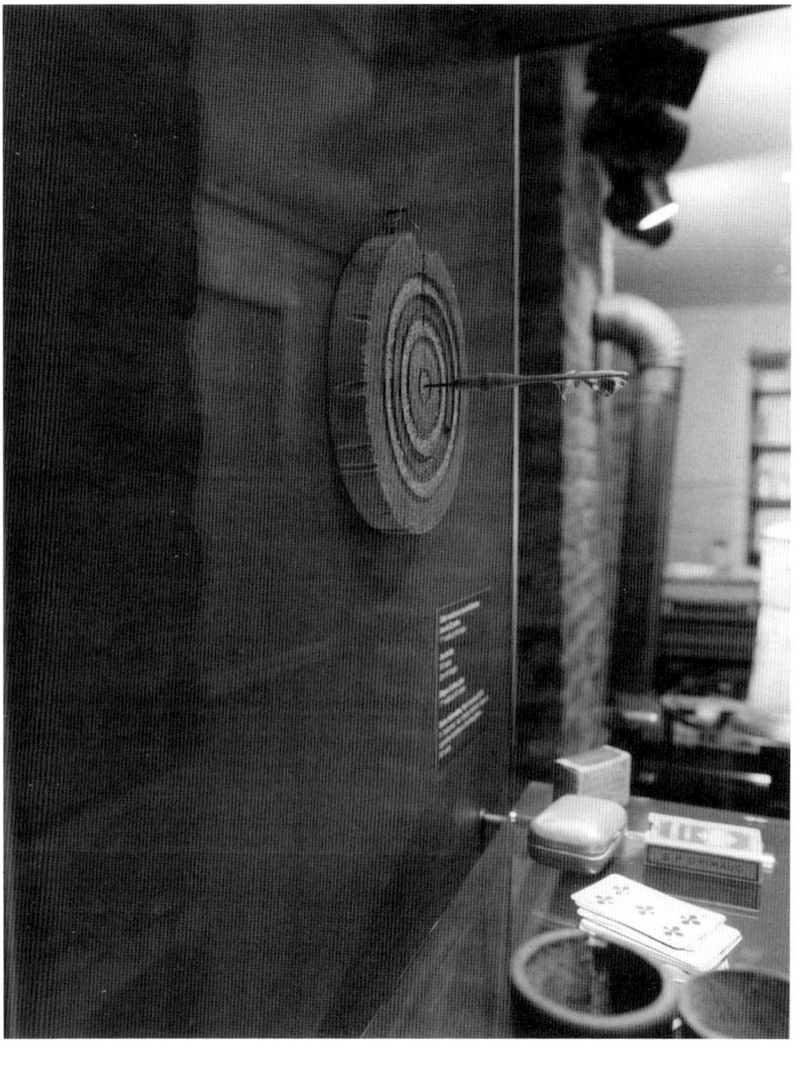

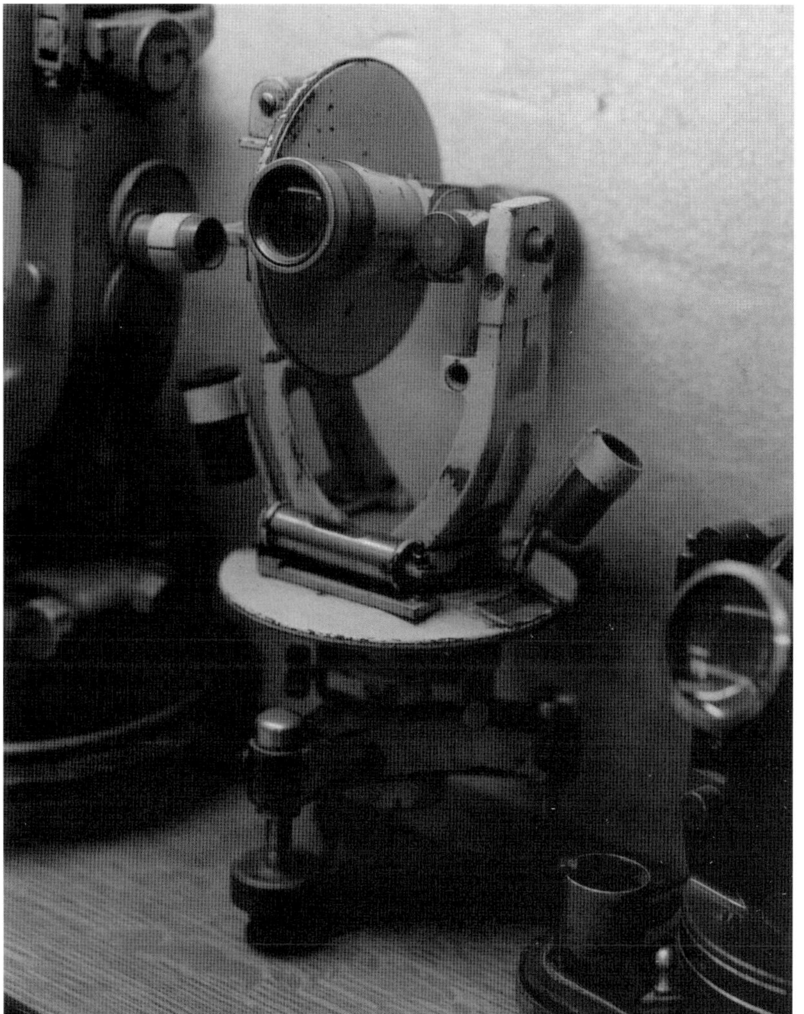

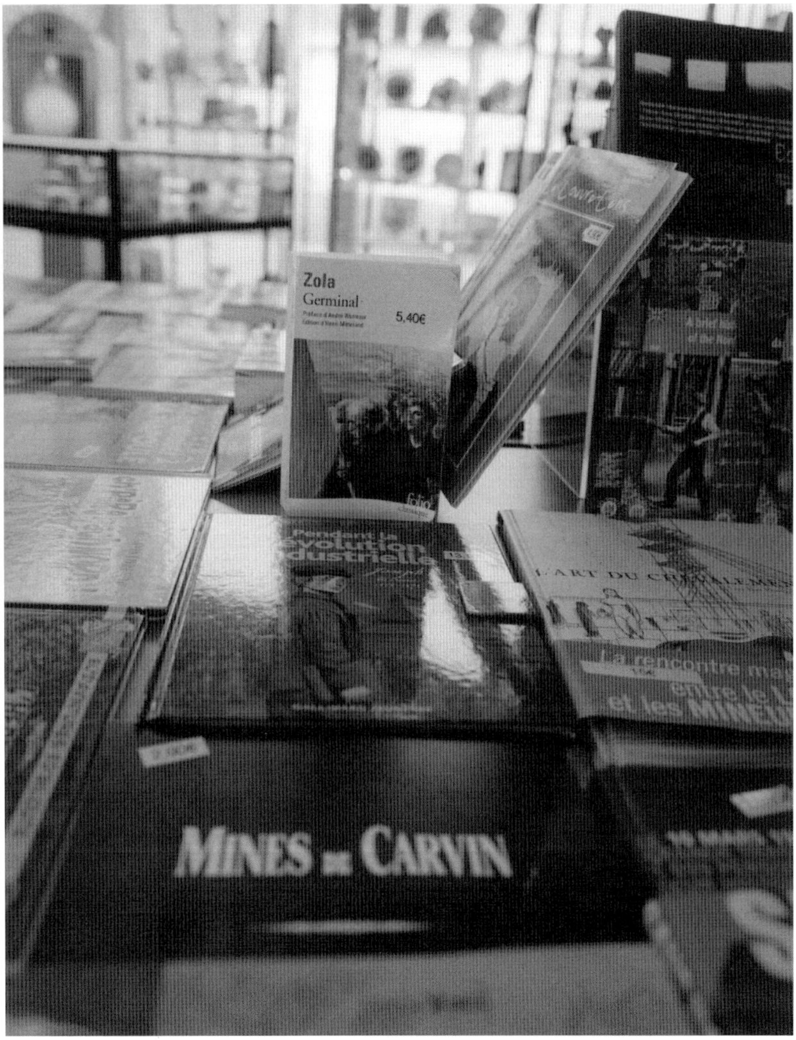
Zola
Germinal
5,40€
folio
MINES
CARVIN

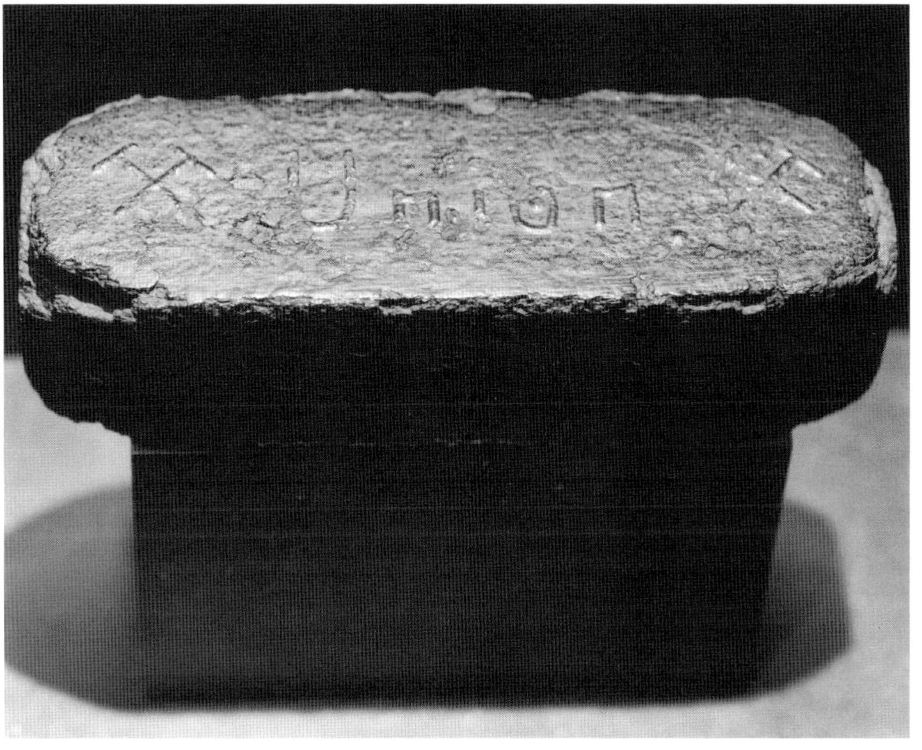

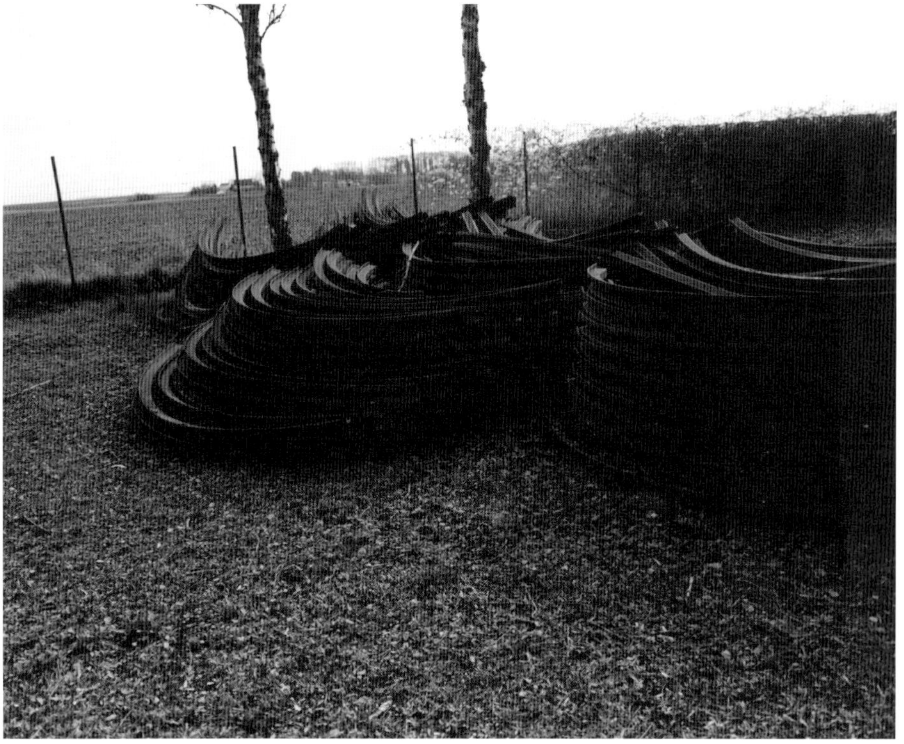

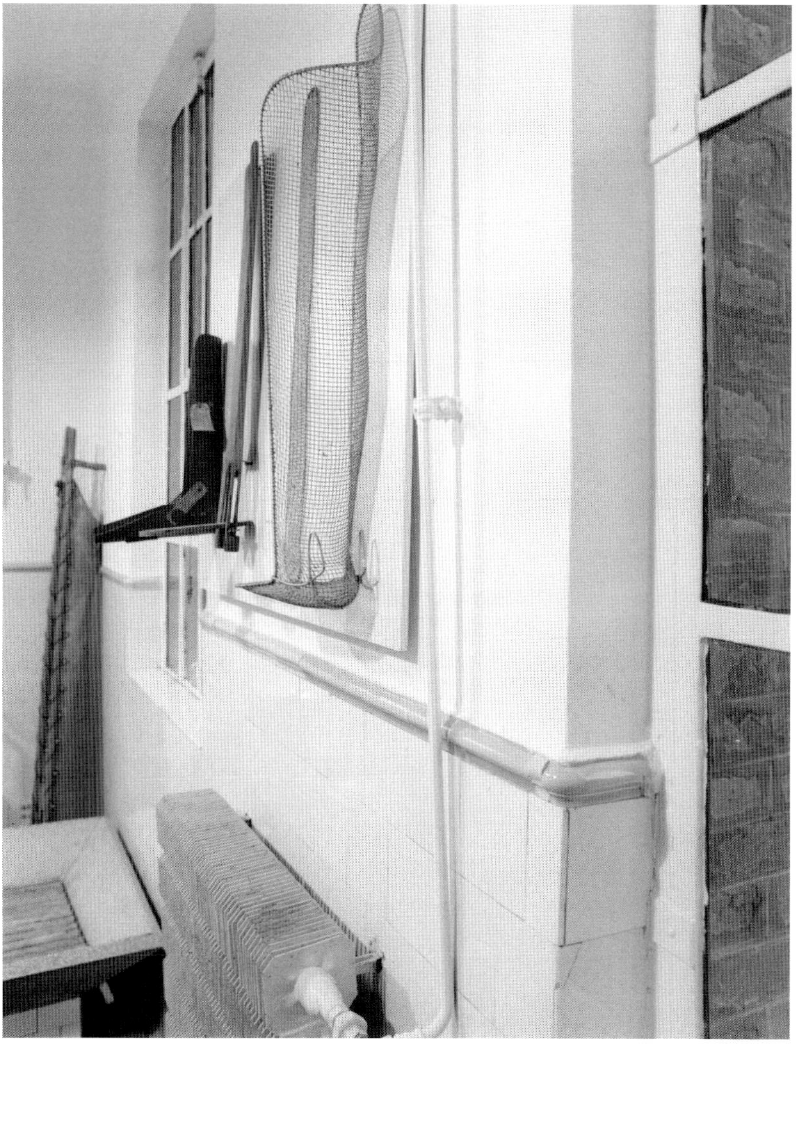

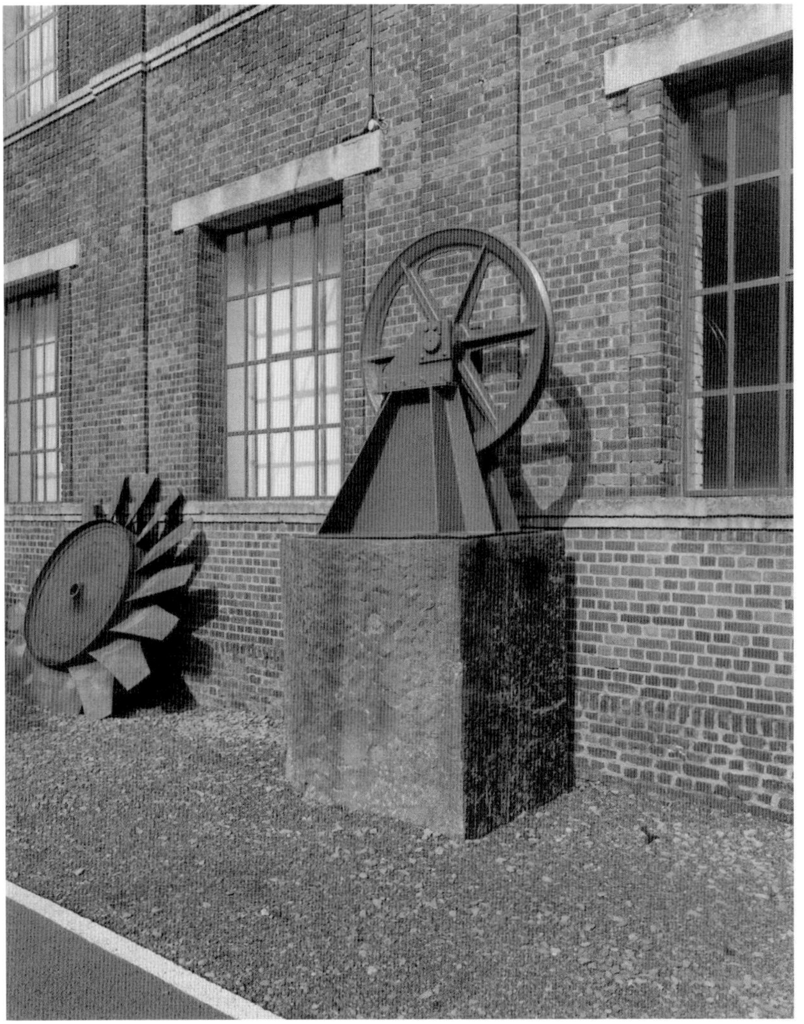

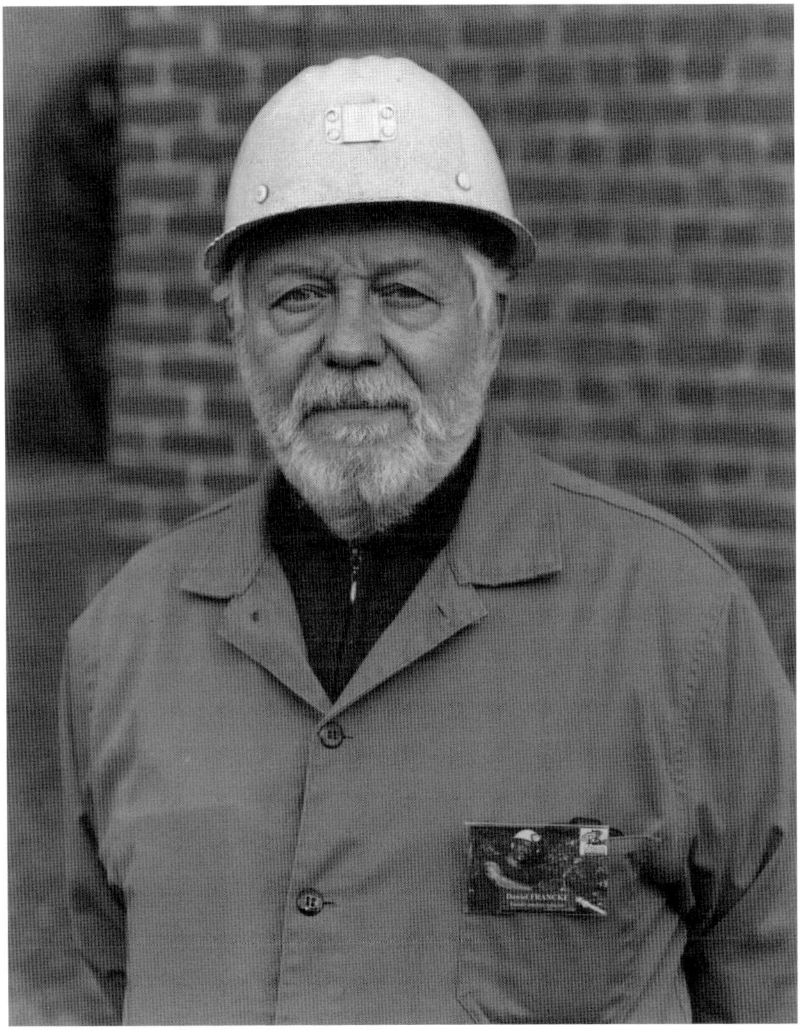

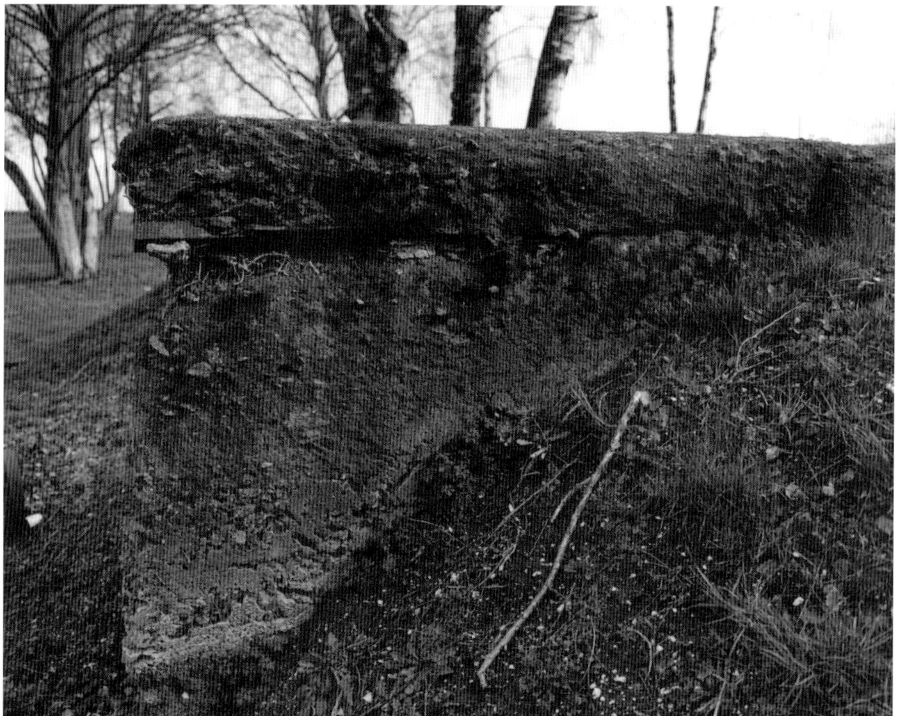

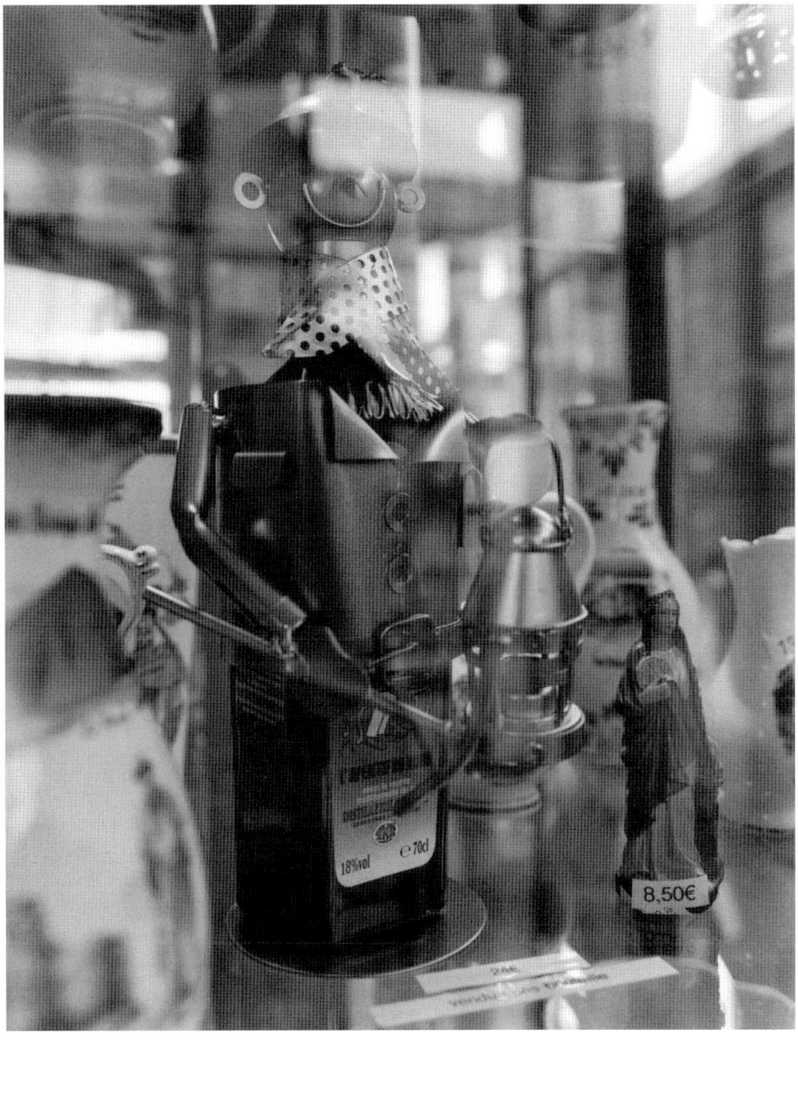
18%vol
e 70cl
8,50€

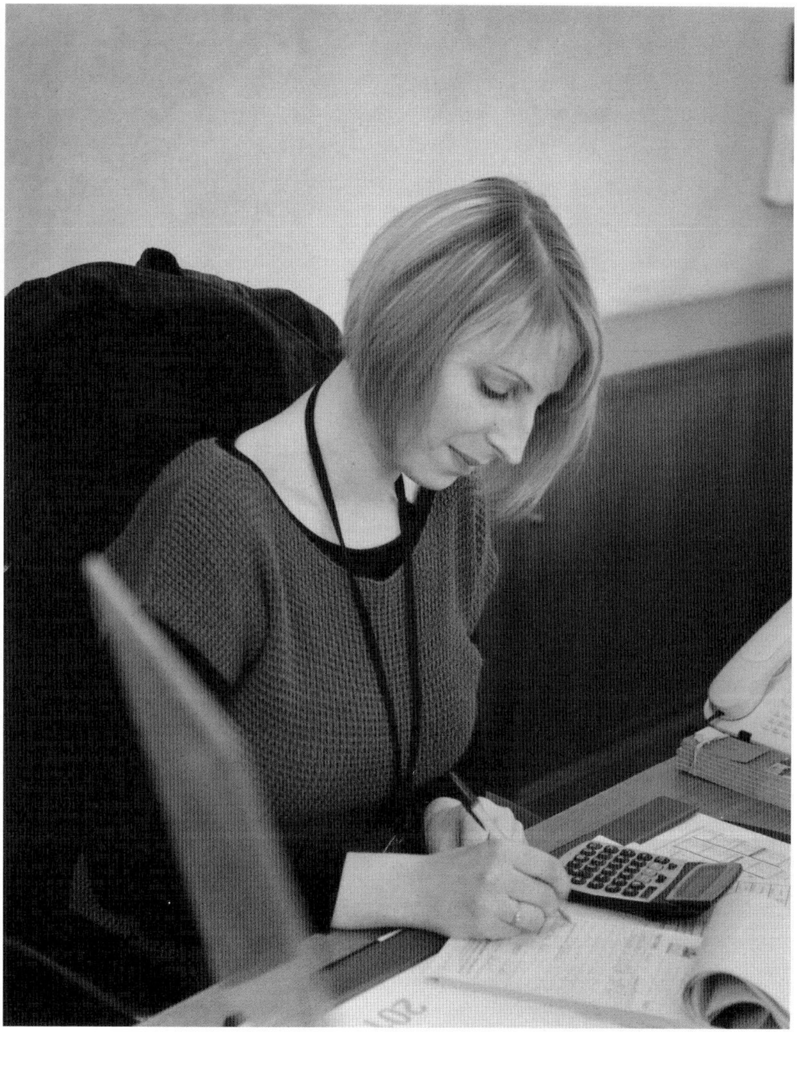

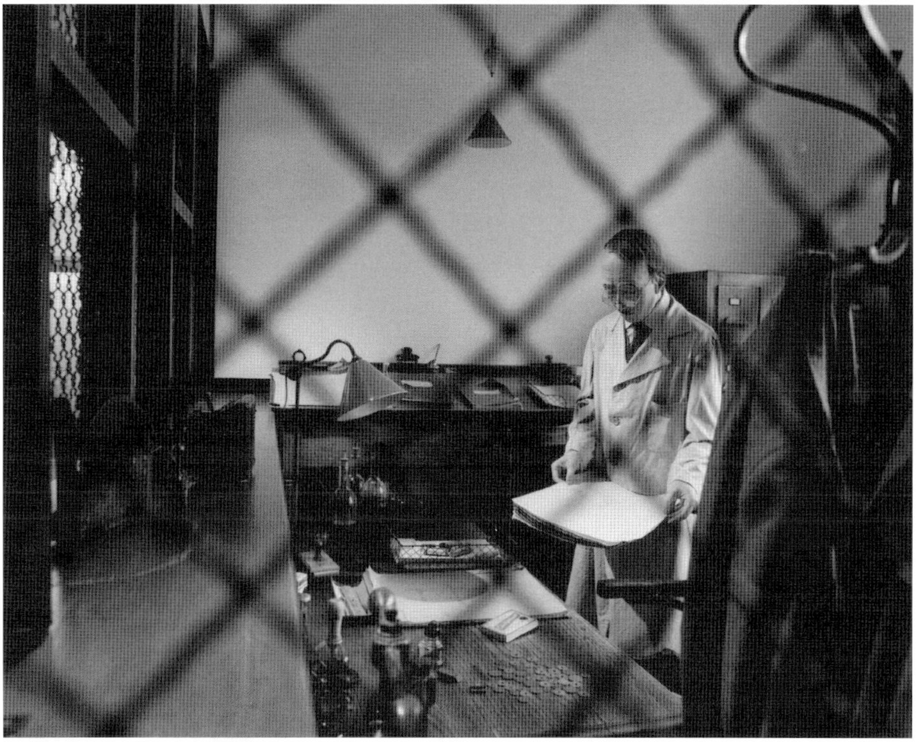

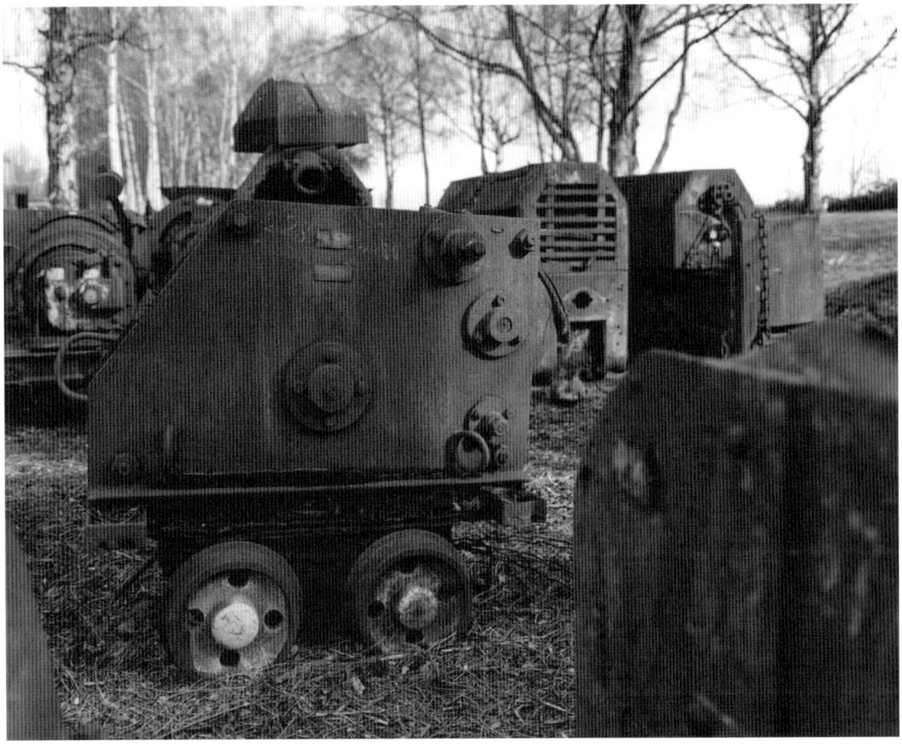

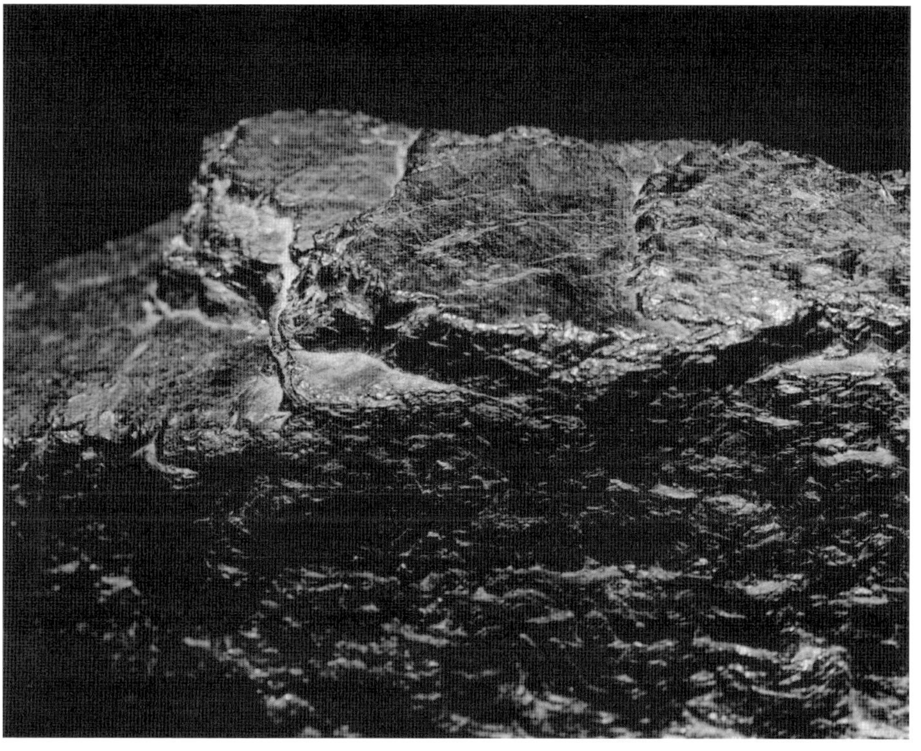

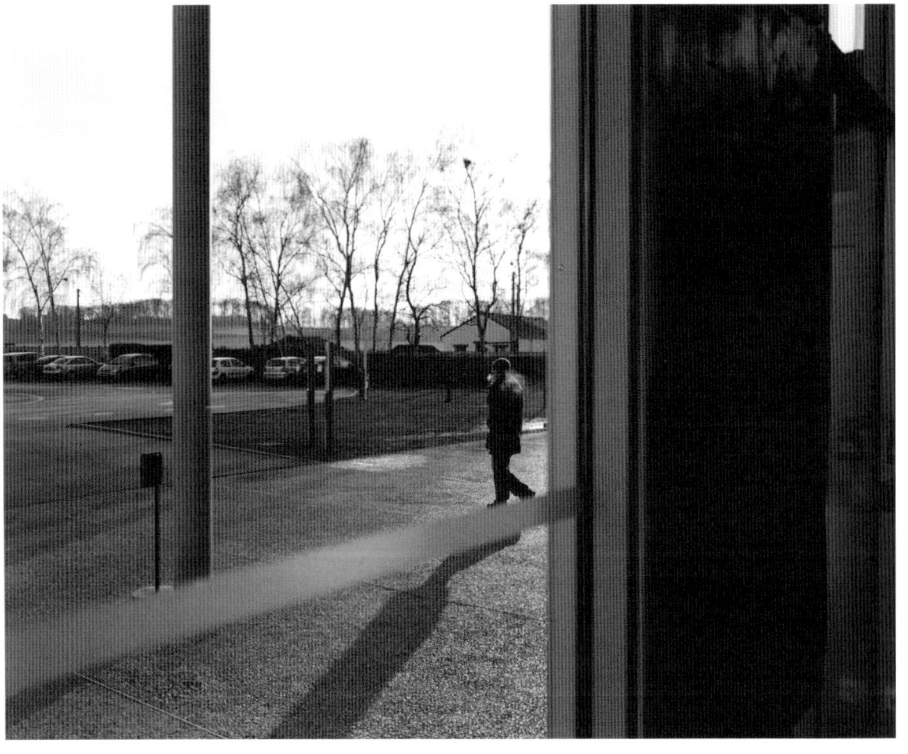

Becoming Heritage

The Maison Syndicale du Minier in Lens was the main mining trade union institution in the region. It was constituted after a big mining catastrophe in 1906, the "catastrophe de Courrières" with 1099 deaths, which impulsed the mining trade-unionism in the region. The building was finished in 1911 but it was heavily damaged during World War I and reconstructed in 1923. It was linked to the CGT and was the main organizational space for mineworkers, including strikes and demonstrations, during the period when the mines were active in the region. It also housed printing facilities where most of the trade union publications and printed matter were produced. The last renovation is from 1994 and two years later it was incorporated into the catalogue of historical monuments. Since the 1990s it is the home of the association Mémoires et Cultures de la région minière. It is currently the historical archive of the coal-workers' organizations and mining trade union activities. It holds an exhibition space and is also the site of the organisation Pays d'Art et d'Histoire de Lens-Liévin. The photographs include various details of the building interior, including the administration council meeting room, the former printing facilities with piles of old trade-union printed matter, and the offices during the inventory work of one of the collections, where a small Lenin bust is used as support for the photographing of the objects.

Also in Lens, the regional Nord-Pas-de-Calais administration started in 2010 a new "identitarian" project called Mineurs du Monde, as a response to the new urban dynamic determined by the opening of the Louvre-Lens and the UNESCO world heritage minefield declaration. It is an attempt to contribute to the various ongoing processes for the preservation and promotion of historical memory of mining labor, to also create a positive public opinion for such historical re-signification of mining labor and to establish an international network of historic minefields turned into cultural sites as part of the global industrial reconversion. Mineurs du Monde is not linked to any specific site but is defined by activities organized in different sites of the region. The photographs were made in one of the events promoted by the organization, specifically a lecture and screening series organized at the Faculté

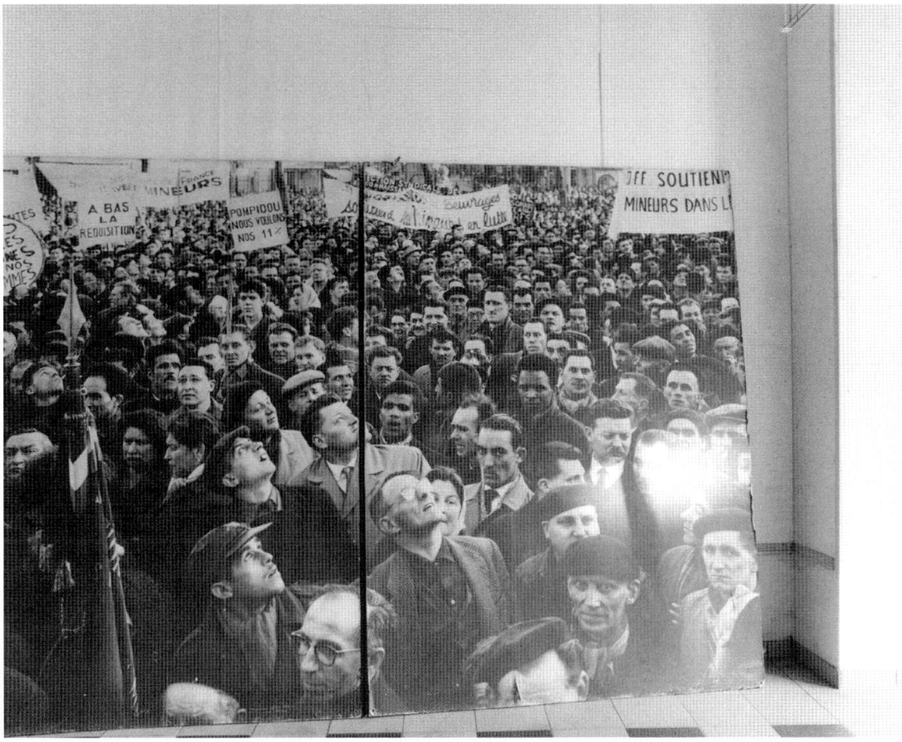
MINEURS
A BAS
LA
REQUISITION
POMPIDOU
NOUS VOULONS
NOS 11 %

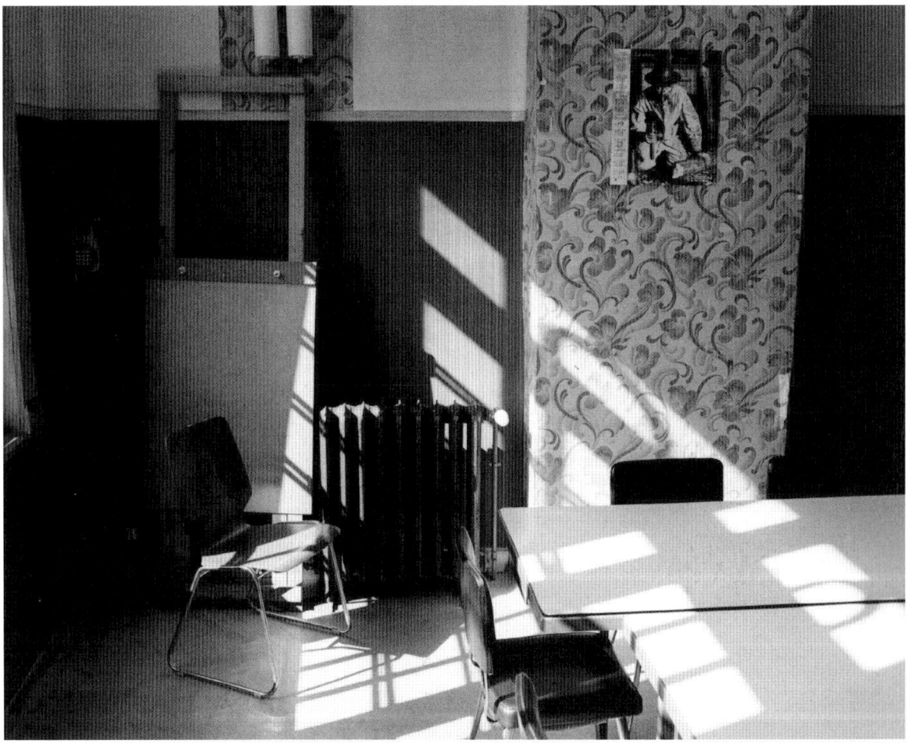

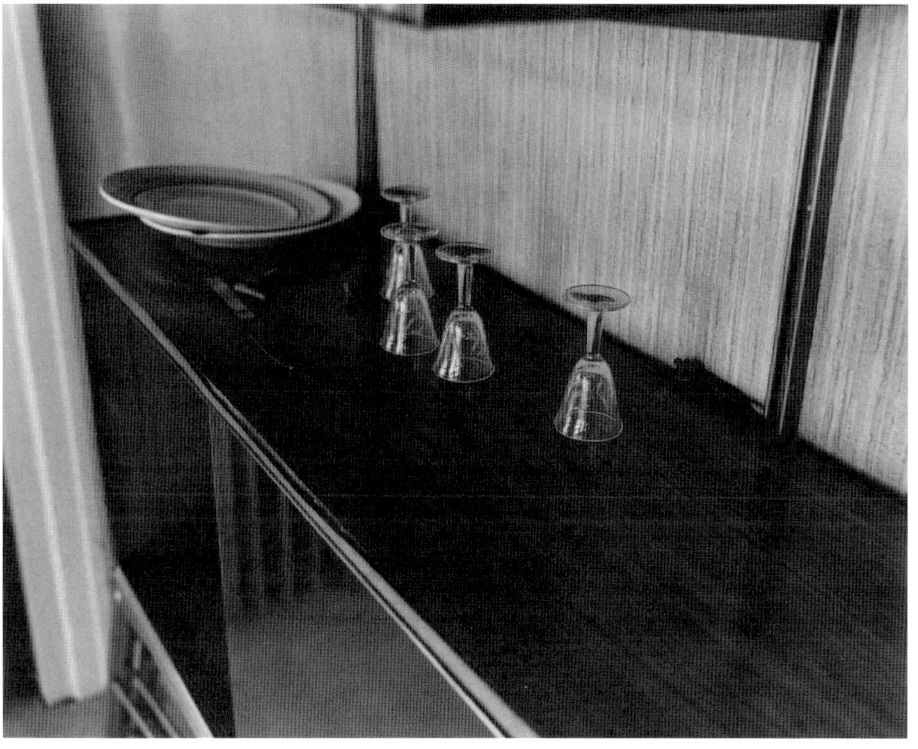

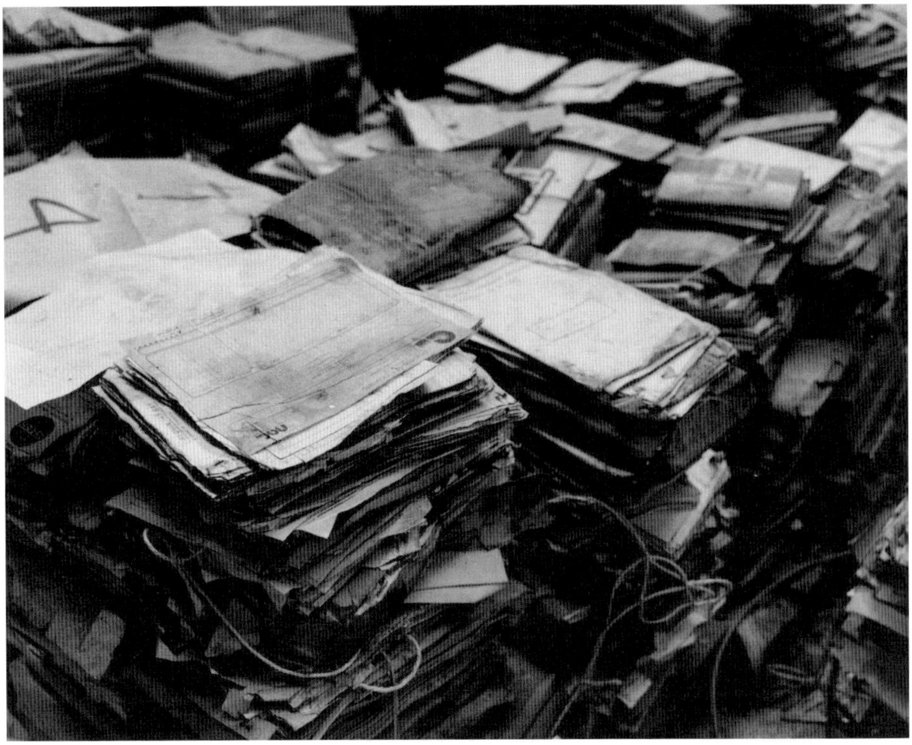

13

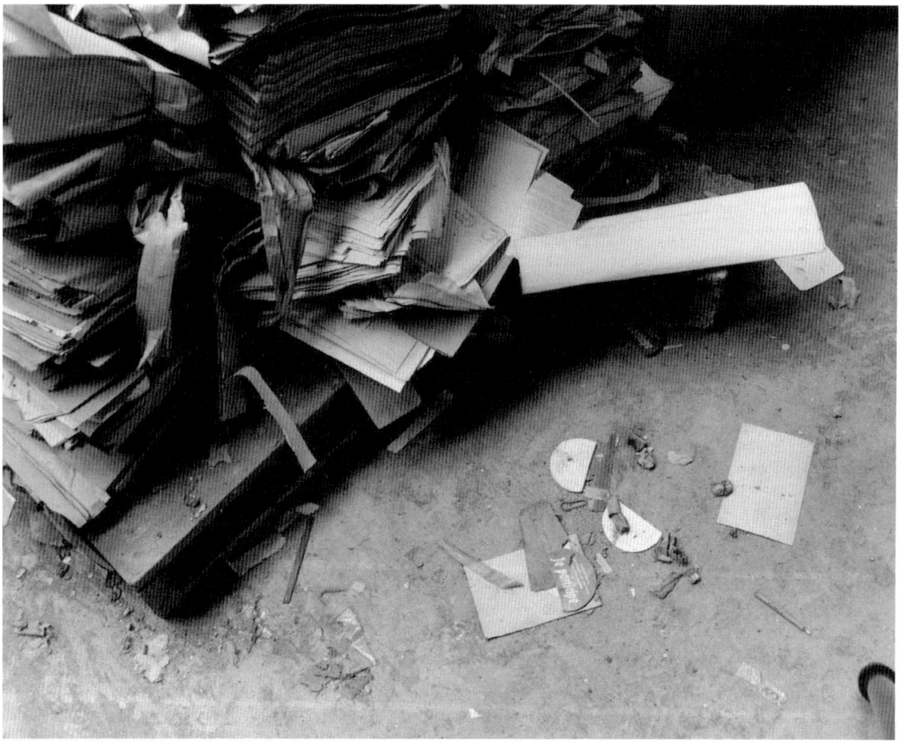

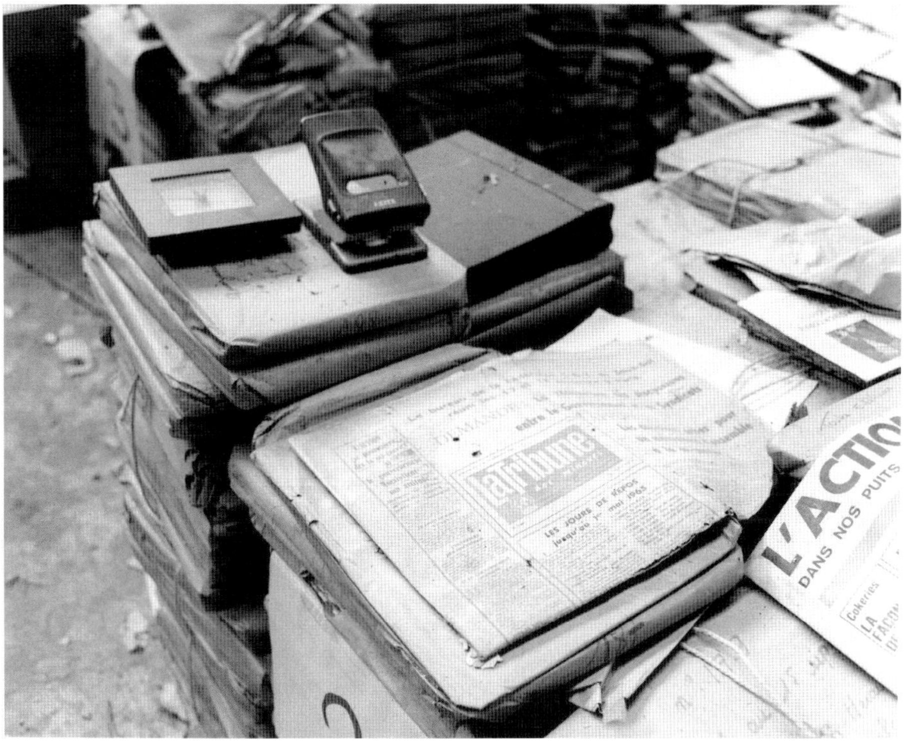
LES JOURS DE REPOS
jusqu'au 1er mai 1965
DANS NOS PUITS
Cokeries
LA

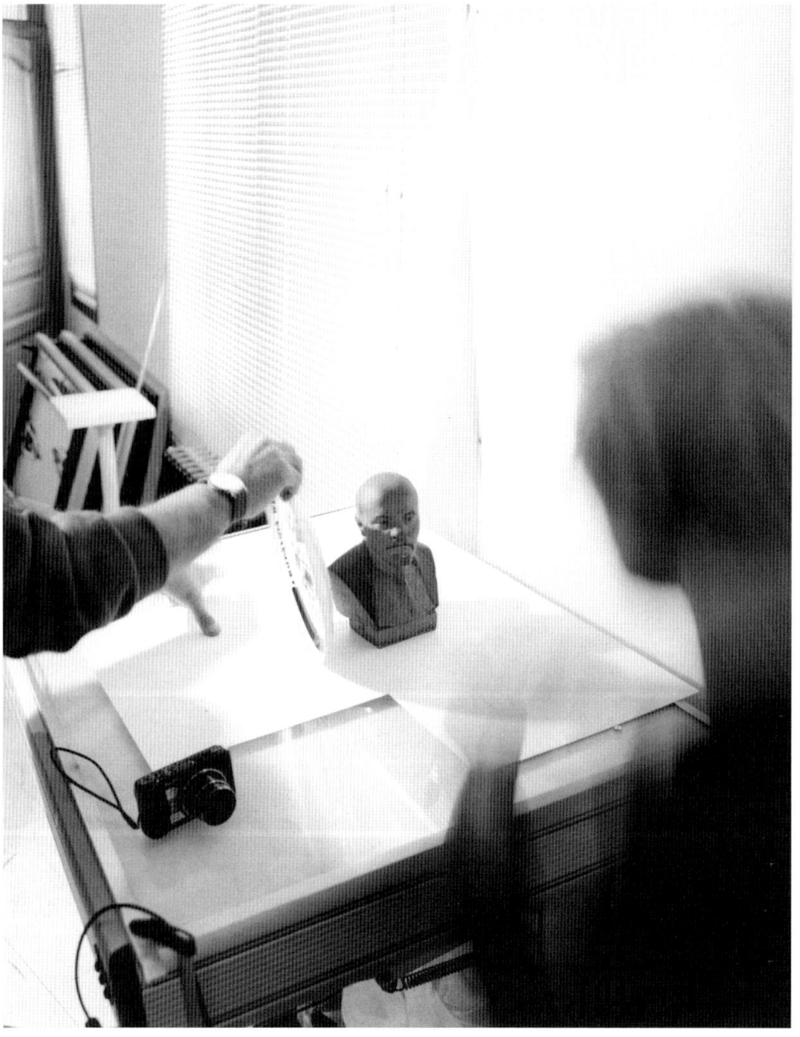

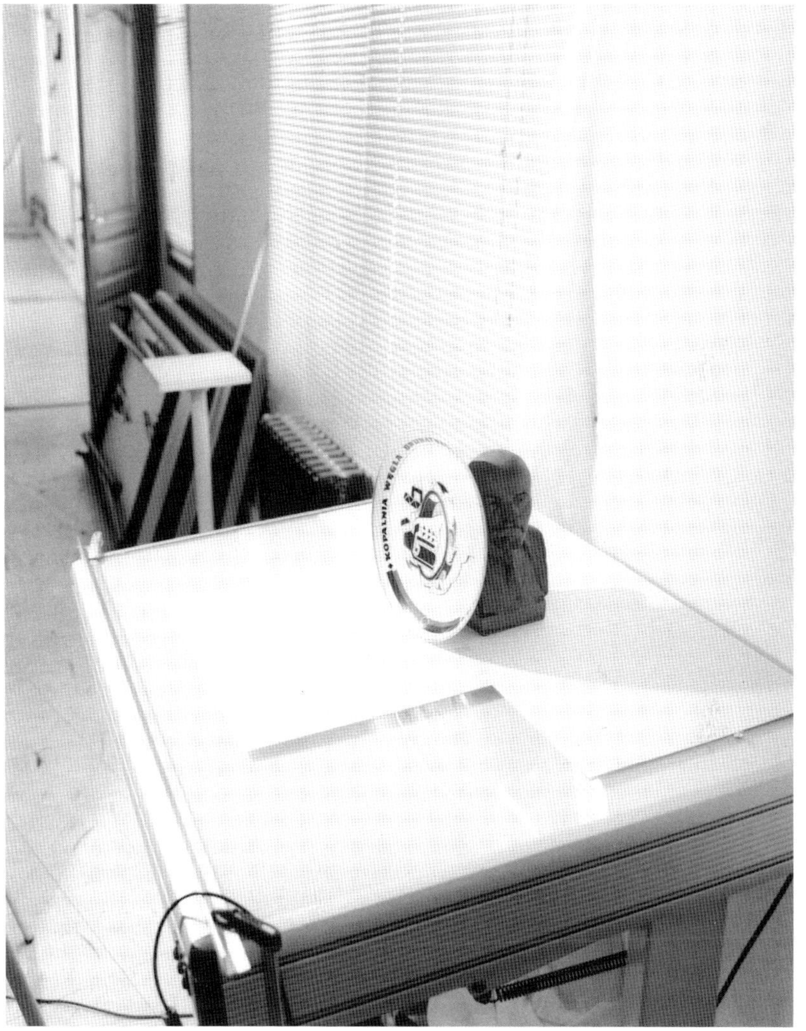
KOPALNIA WEGLA

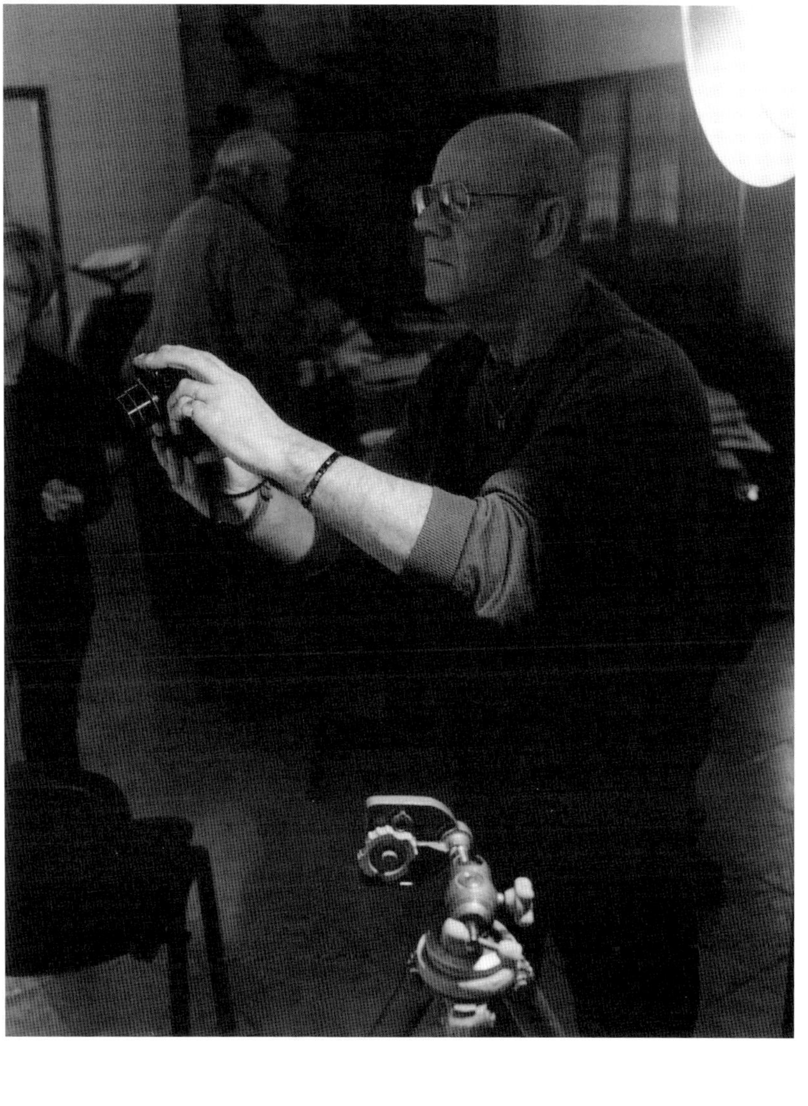

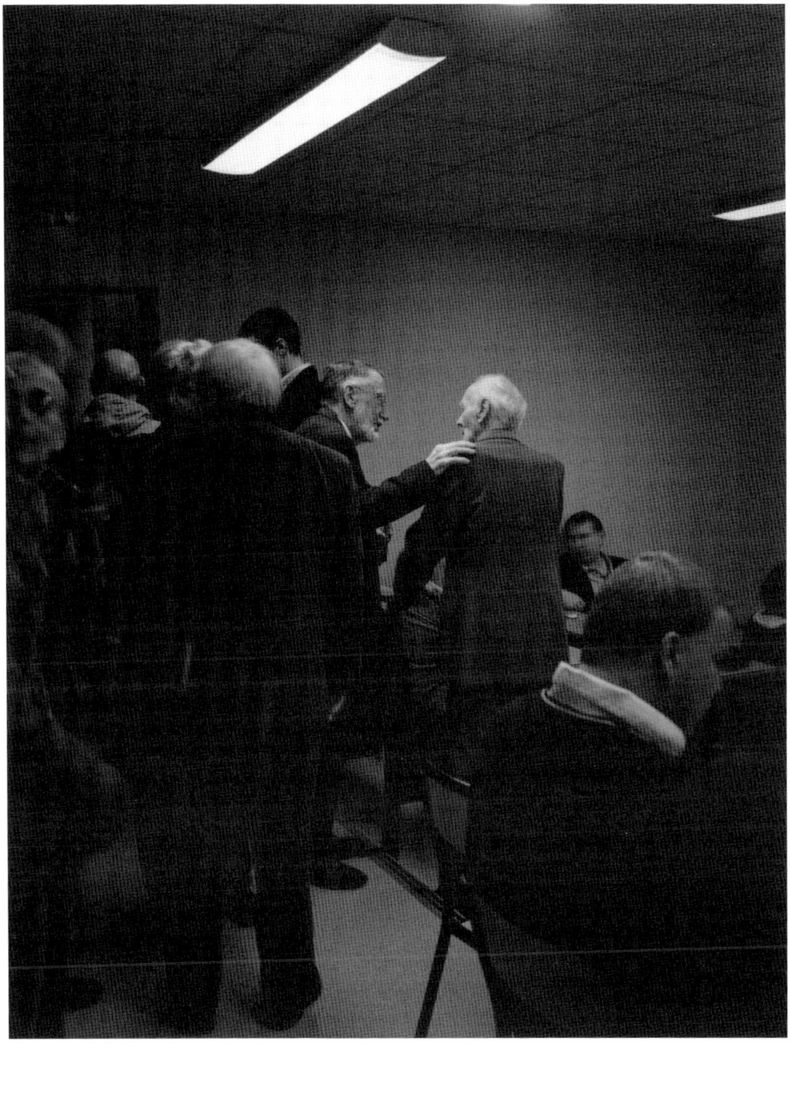

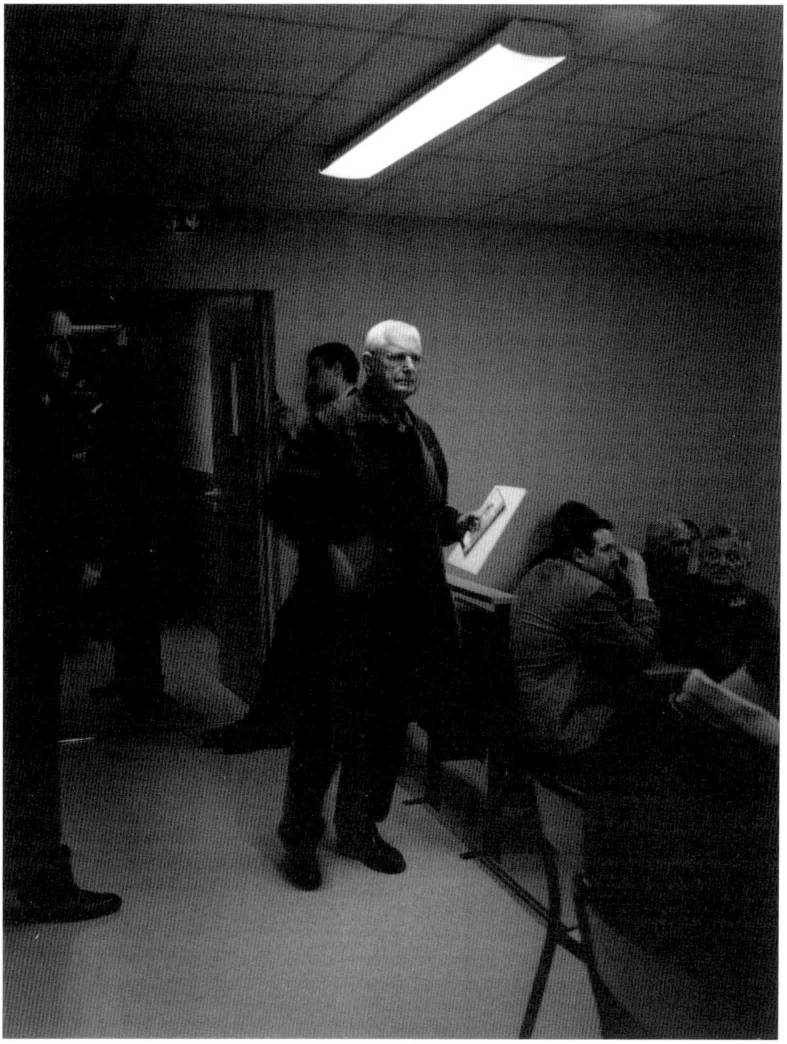

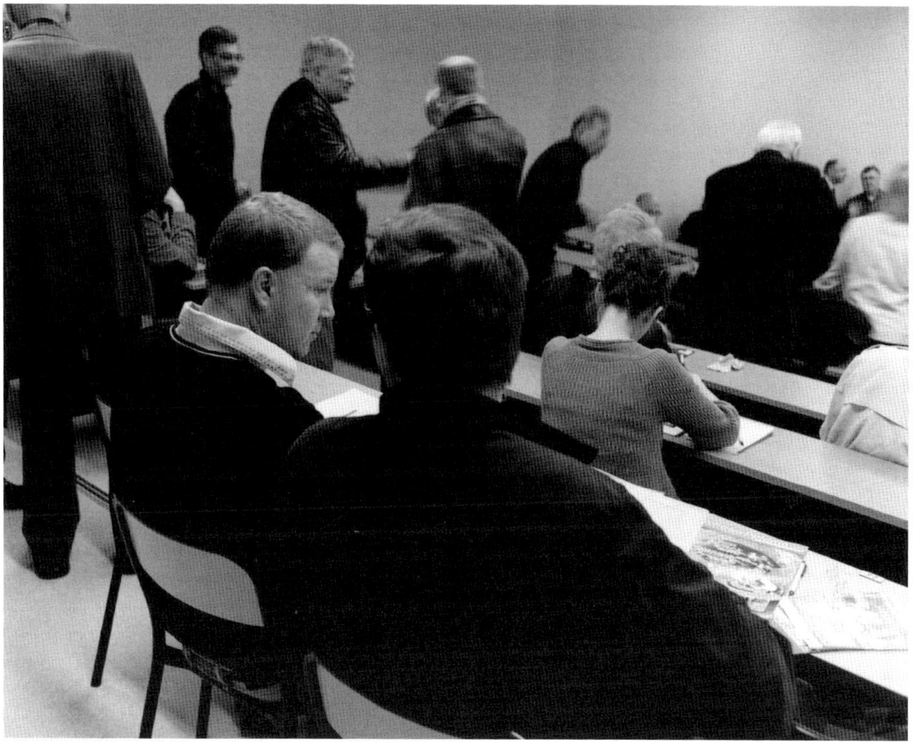

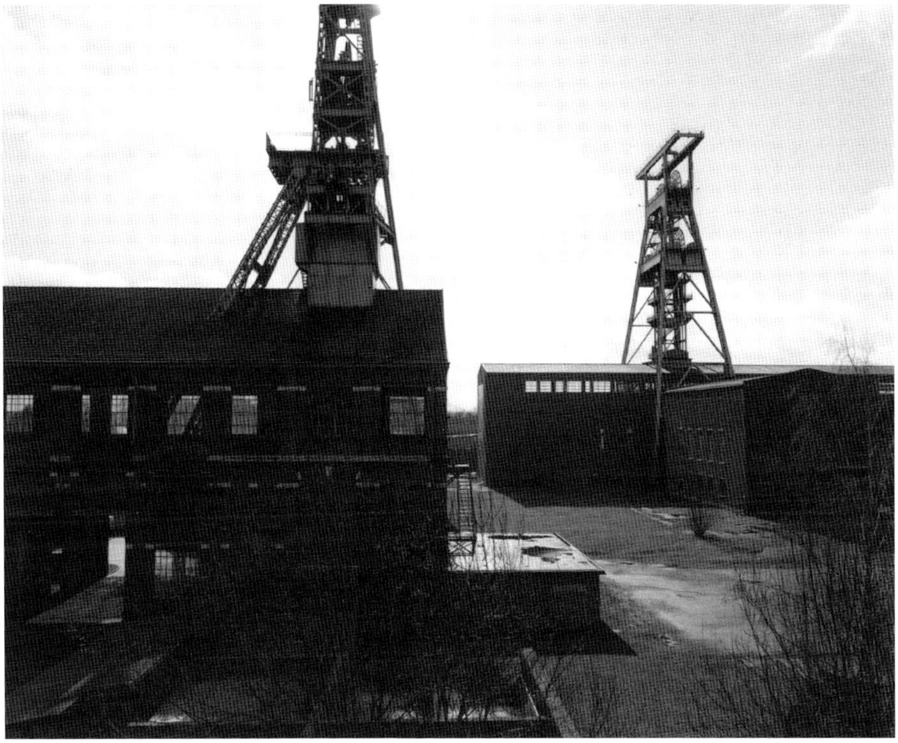

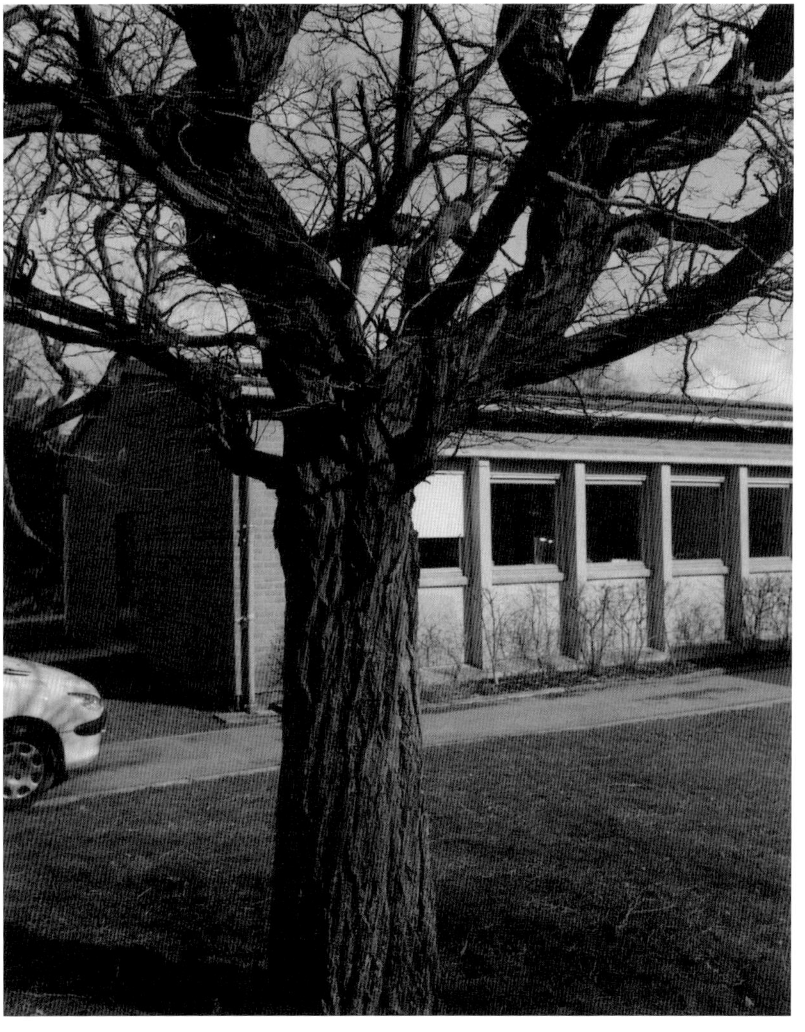

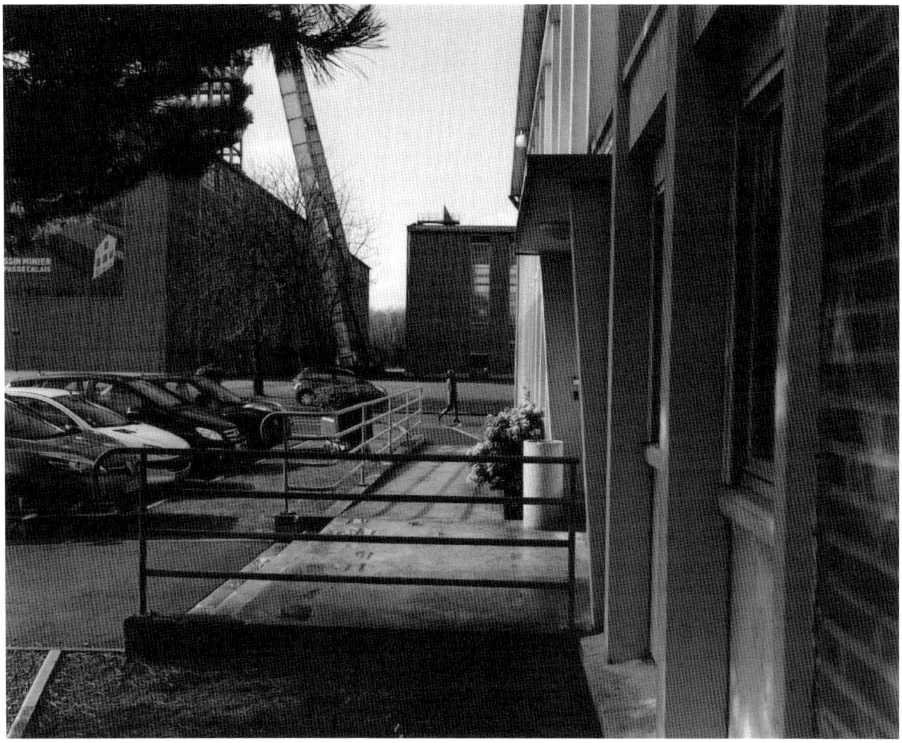
SSIN MINIER
PAS DE CALAIS

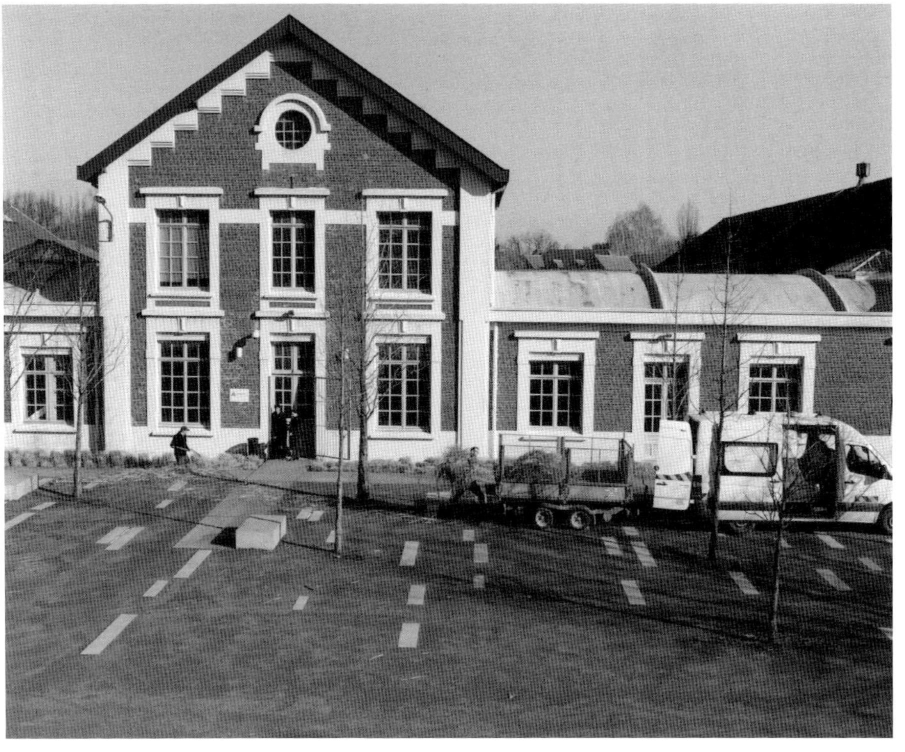

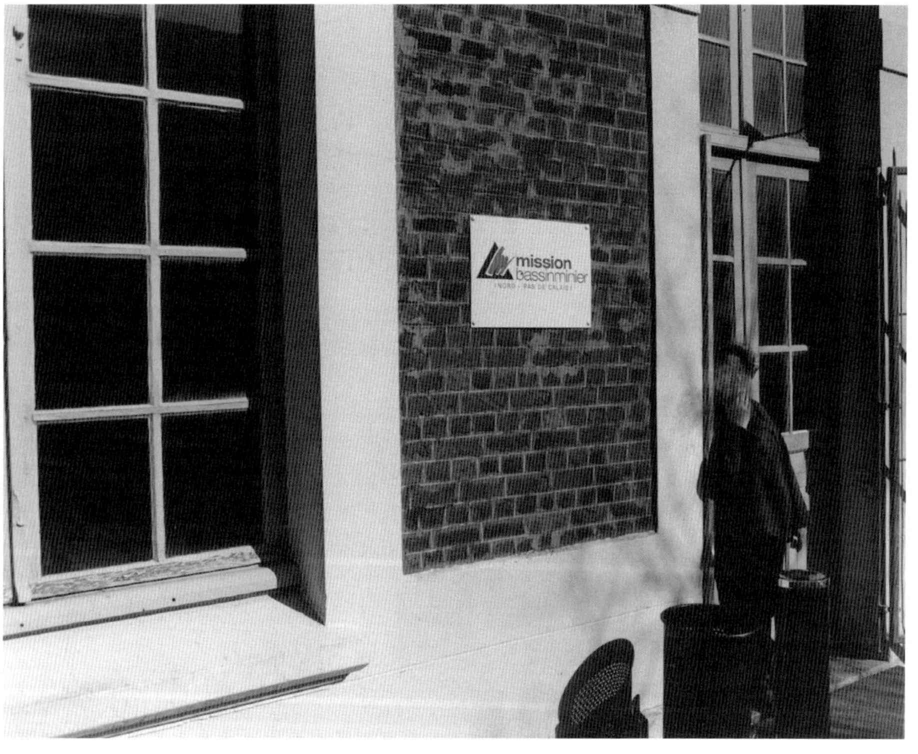
mission
bassinminier

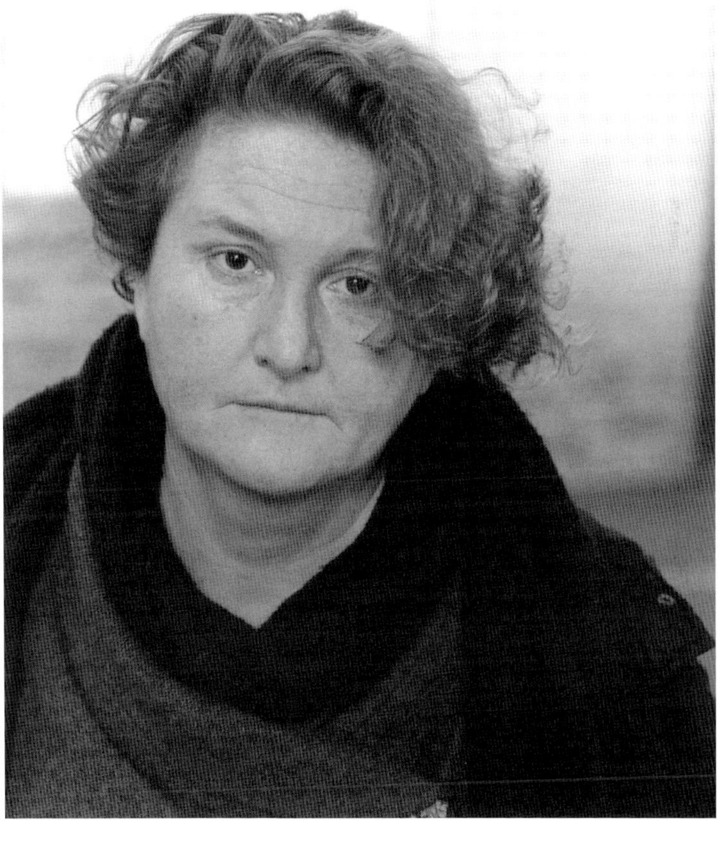

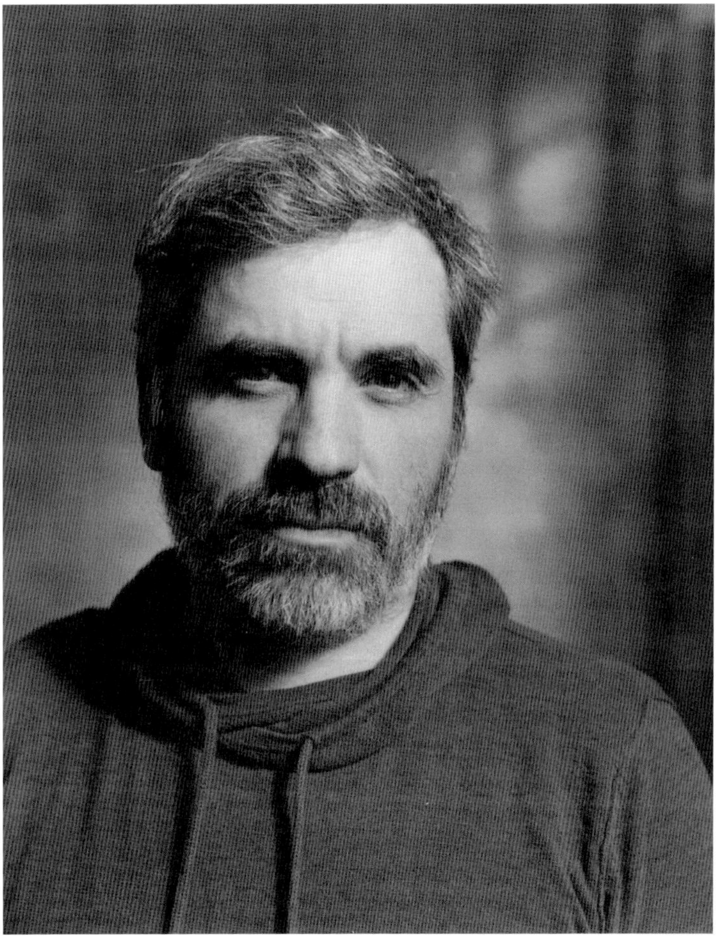

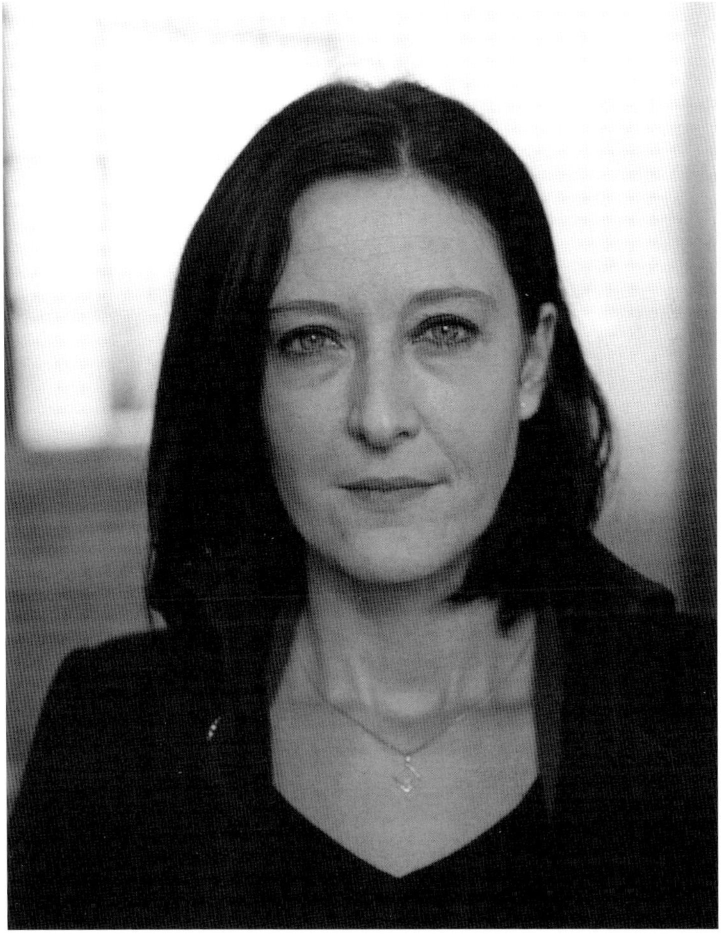

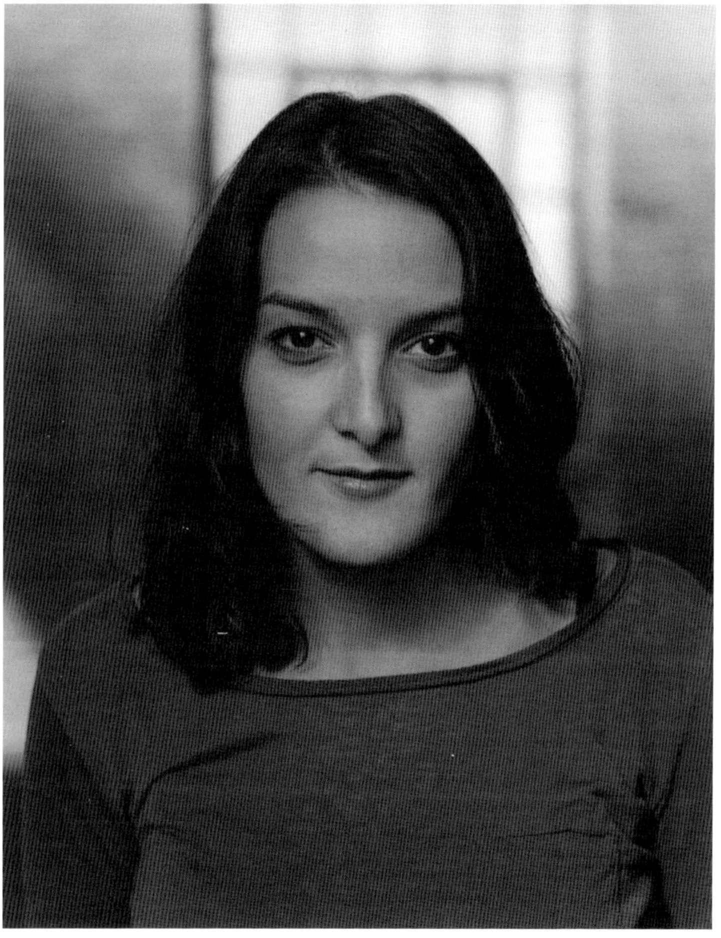

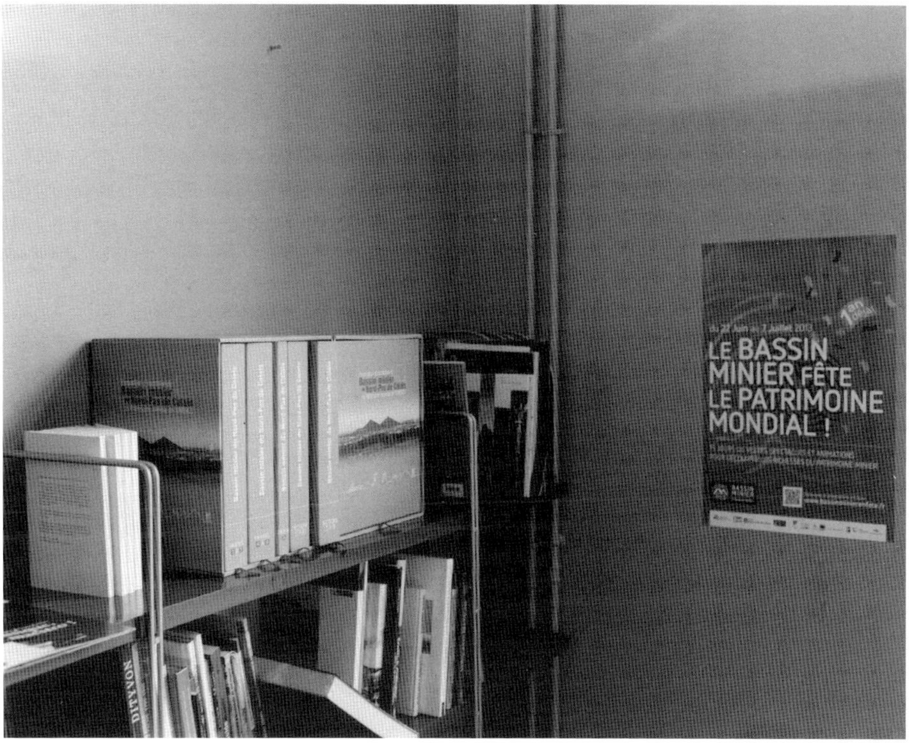
LE BASSIN
MINIER FÊTE
LE PATRIMOINE
MONDIAL !

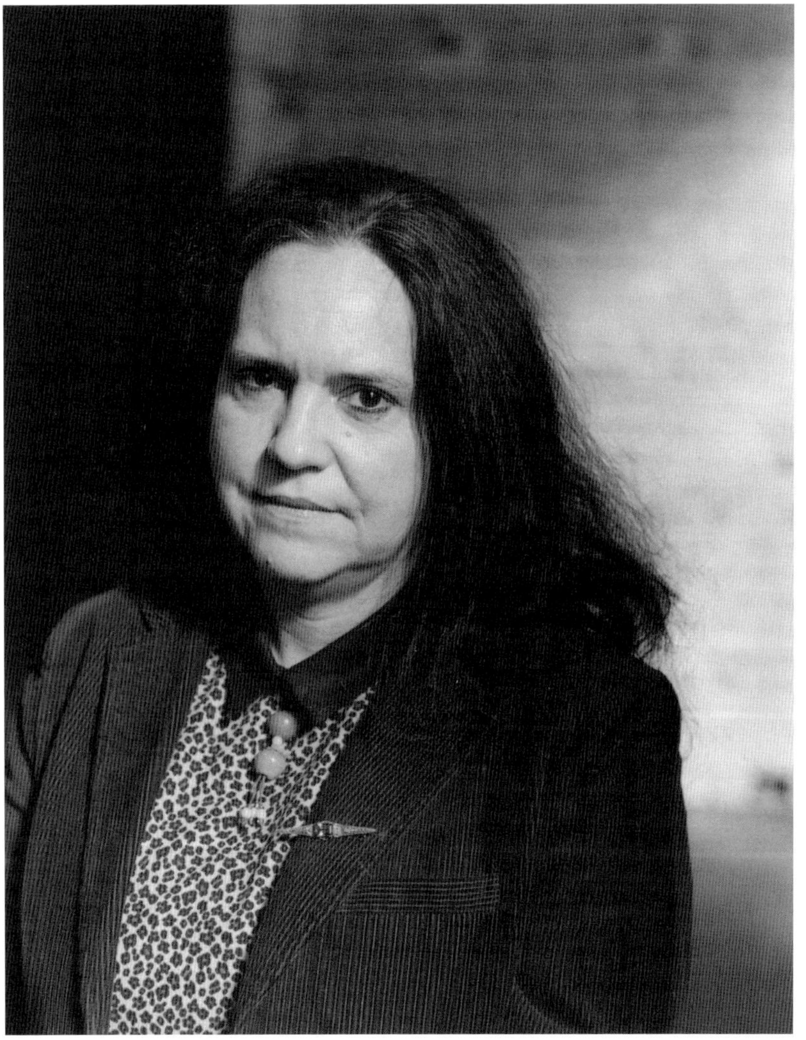

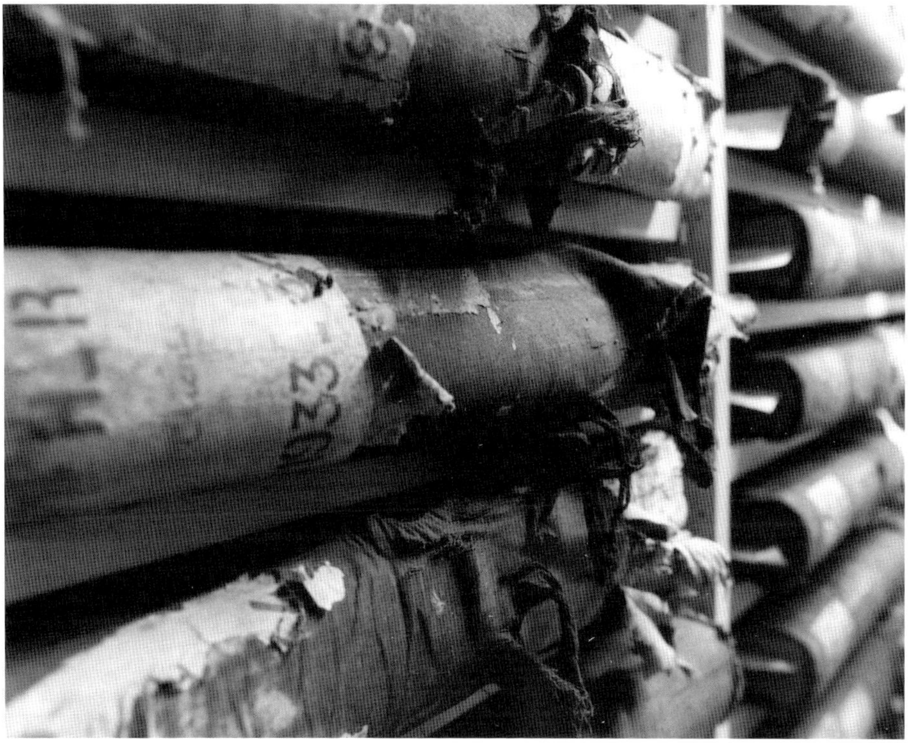

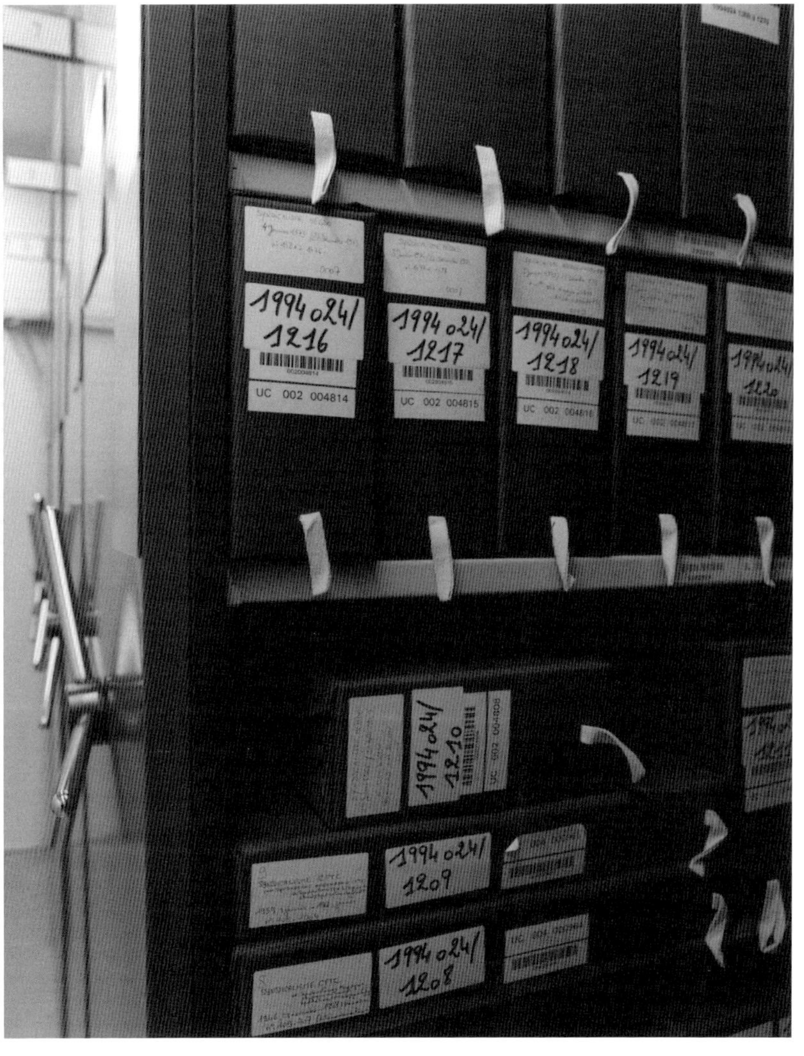
1994024/
1216
UC 002 004814
1994024/
1217
UC 002 004815
1994024/
1218
1994024/
1219
1994024/
1210
1994024/
1209
1994024/
1208

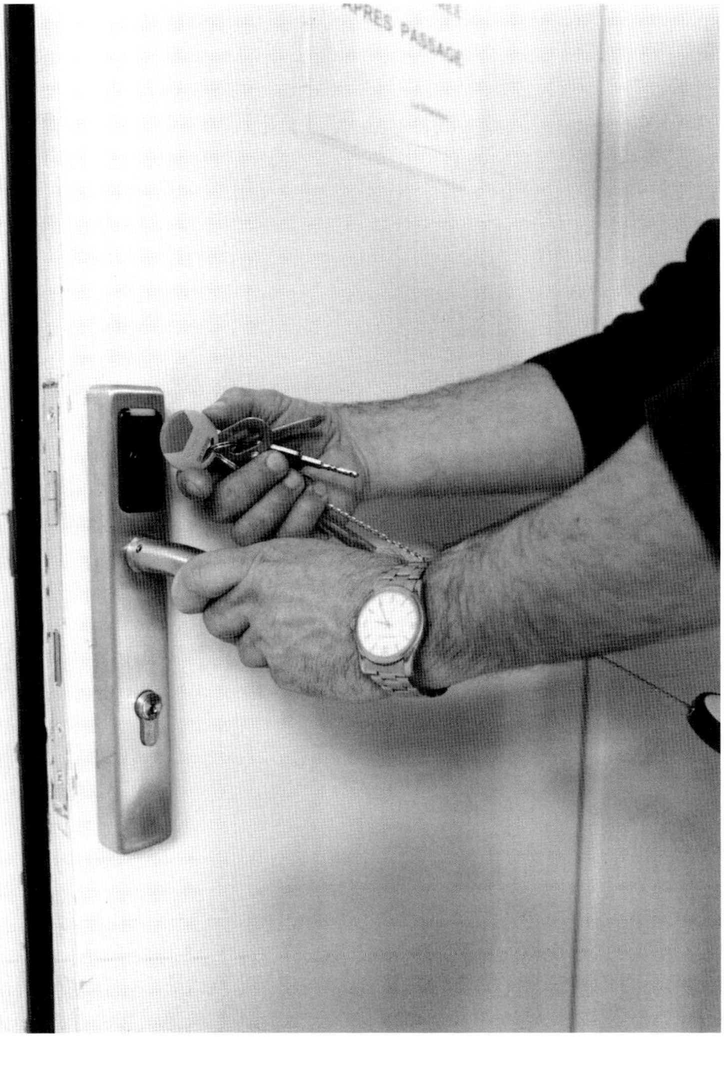
APRES PASSAGE

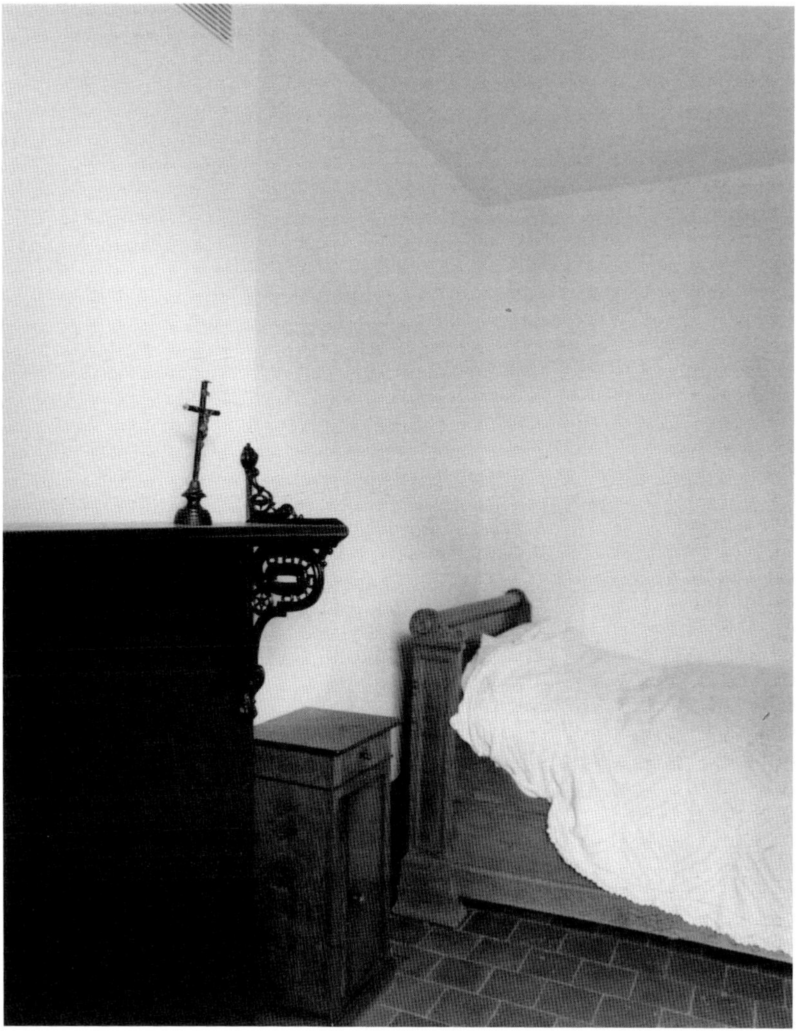

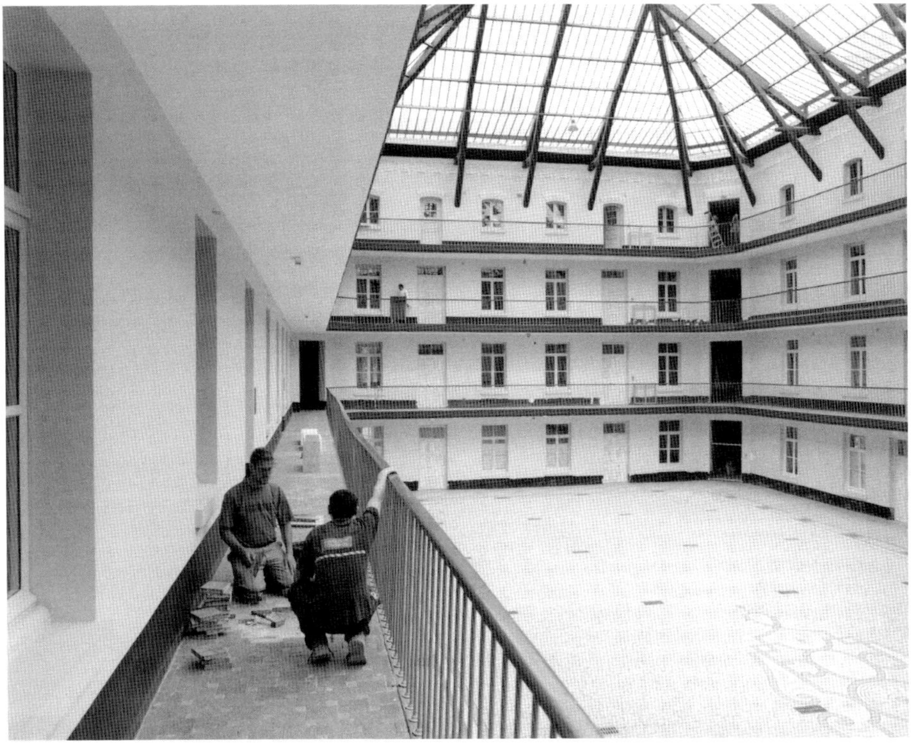

New/Old Economy

New economy enterprises have proliferated in former mining and industrial sites. In Oignies new iron and glass minimalist office buildings coexist with the old mining towers. In the surroundings of Fosse 9–9bis the former miner families housing areas have been renovated and expand as new upper-class residential areas. The Fosse d'Arenberg in Wallers was the shooting site for the film version of Emile Zola's *Germinal* (Claude Berri, 1993) and has been reconverted in a film center, including a television studio.

La Plaine Images in the Boulevard Descat, Tourcoing, not far from Le Fresnoy, is a new cluster of digital creativity, advertising, communication, video-games, digital applications and image enterprises. In this former industrial complex a public-private fix provides a strangely informal and ultramodern work atmosphere in which labor and leisure become impossible to differentiate.

In Aulnoy-les-Valenciennes, at one of the Université de Valenciennes campuses, the Transalley project, a public-private association of companies and industrial organizations, has established a "technopole", a concentration of various institutional and engineering buildings, including the Institut des Sciences et Techniques de Valenciennes. Translalley is constituted by car and railroad industry corporations, and seeks to contribute to competitive, innovative and sustainable transportation and mobility business.

Simultaneous to such new economy and "immaterial labor" bureaucratic landscapes, classic "men-at-work" industrial forms of production exist. Industrial and post-industrial labor overlap. In Crespin, the factory Bombardier Transport SAS is one of the various ones in the region that continues producing trains. The photographs show the assembly line for the train coach fabrication (metal work, furnish, finishing, etc.).

The Bouchain Power Plant or Centre de Production Thermique was established in 1970 and had originally two production units, one of them stopped in 1995. It belongs to the EDF energy group. It is one of the few remaining coal-based power stations in Northern France and in 2014, during the shooting, it was actually in the process of turning into gas. The transition is scheduled for 2015. The coal arrives from

Russia, Australia or South Africa by ship to France and then by train to Bouchain. Then it is stored, processed and burned to produce energy, which is the sequence the photographs follow. The big cooling tower of the plant looks like those from nuclear plants, which makes it unique among coal power plants in France.

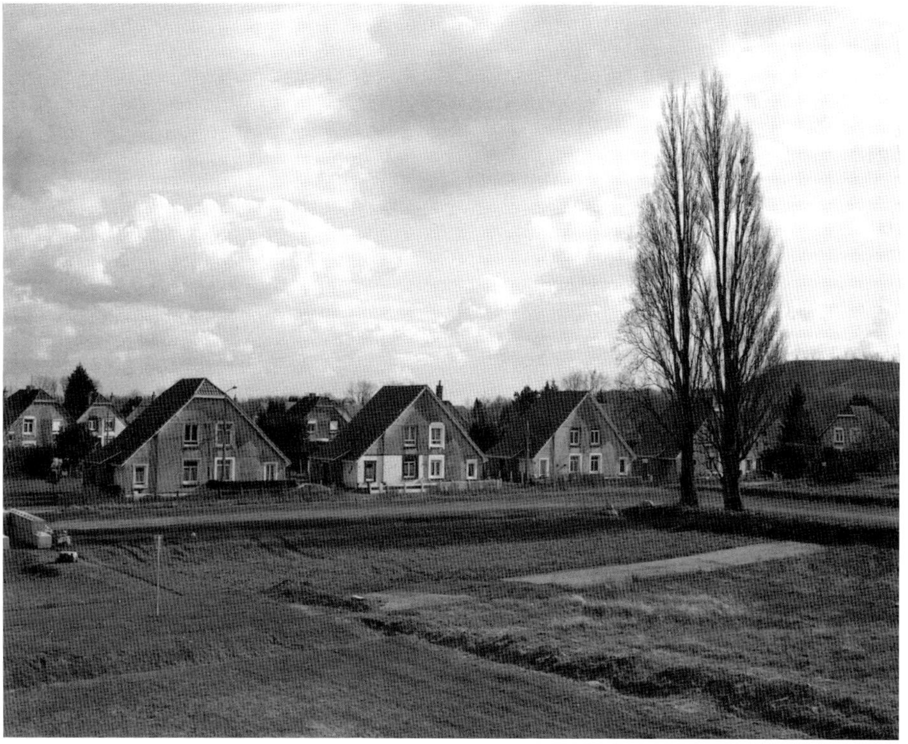

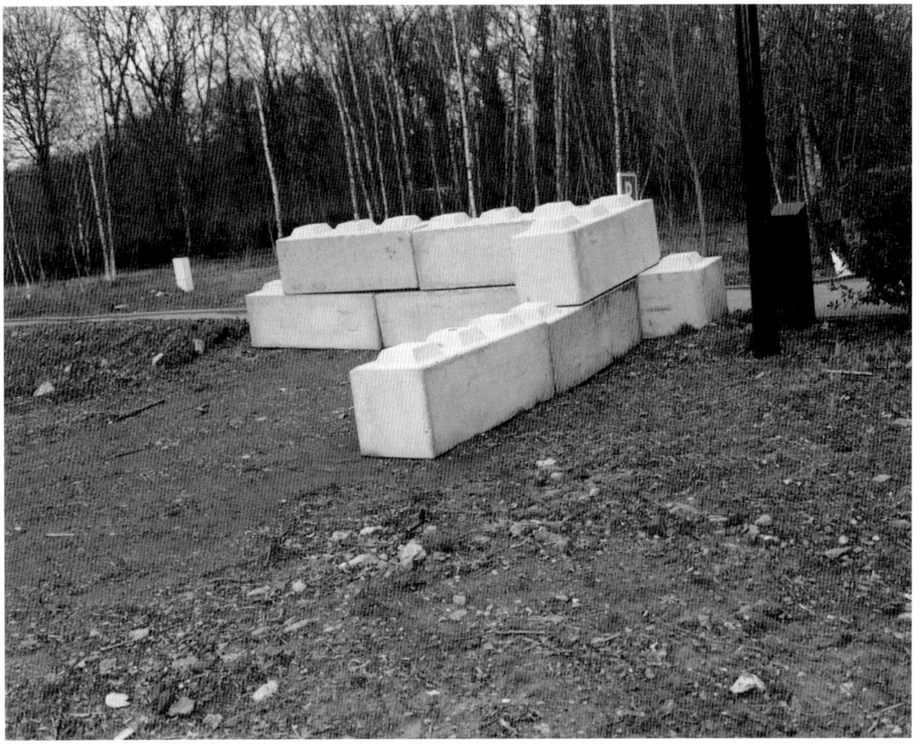

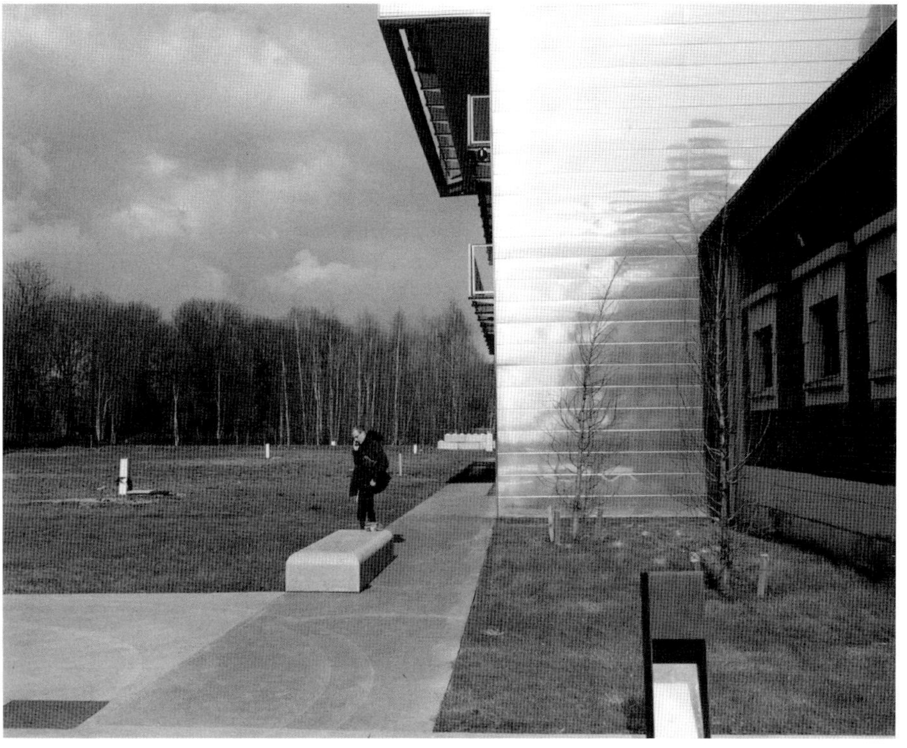

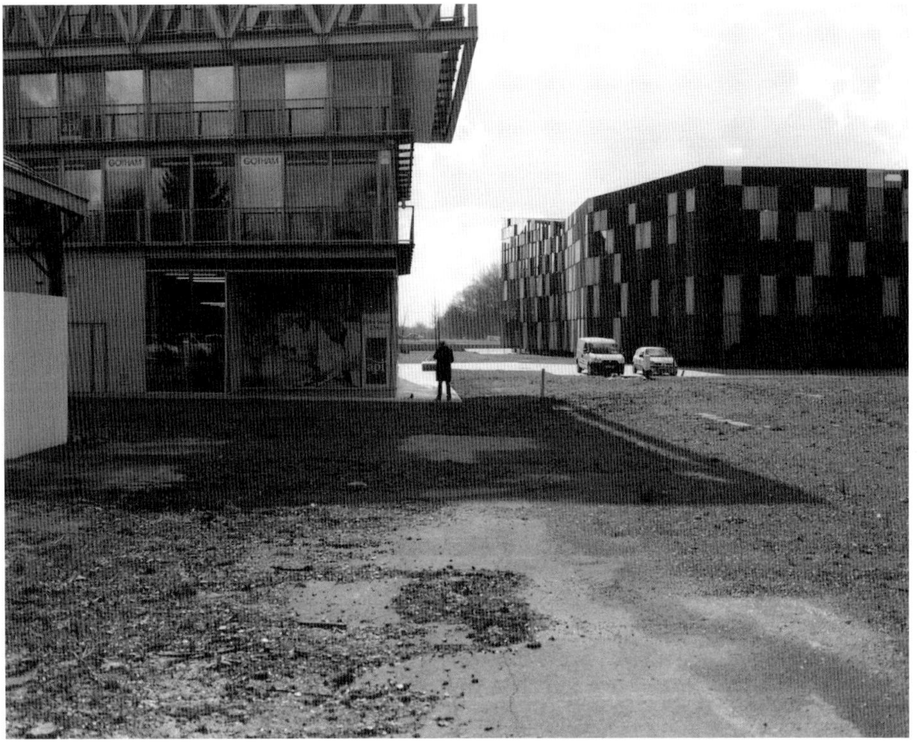

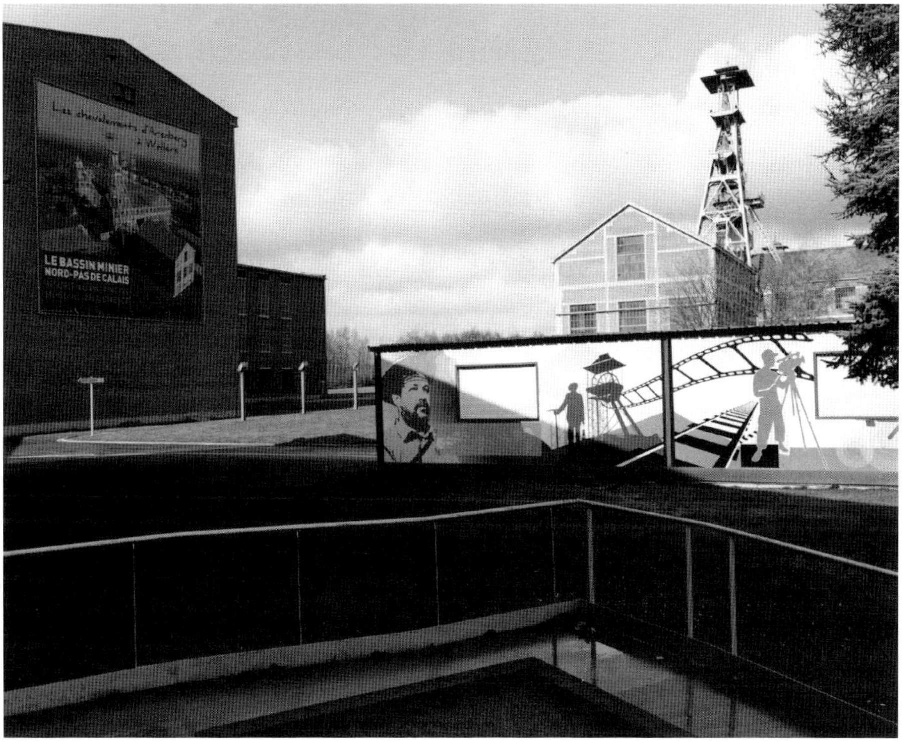
LE BASSIN MINIER
NORD-PAS DE CALAIS

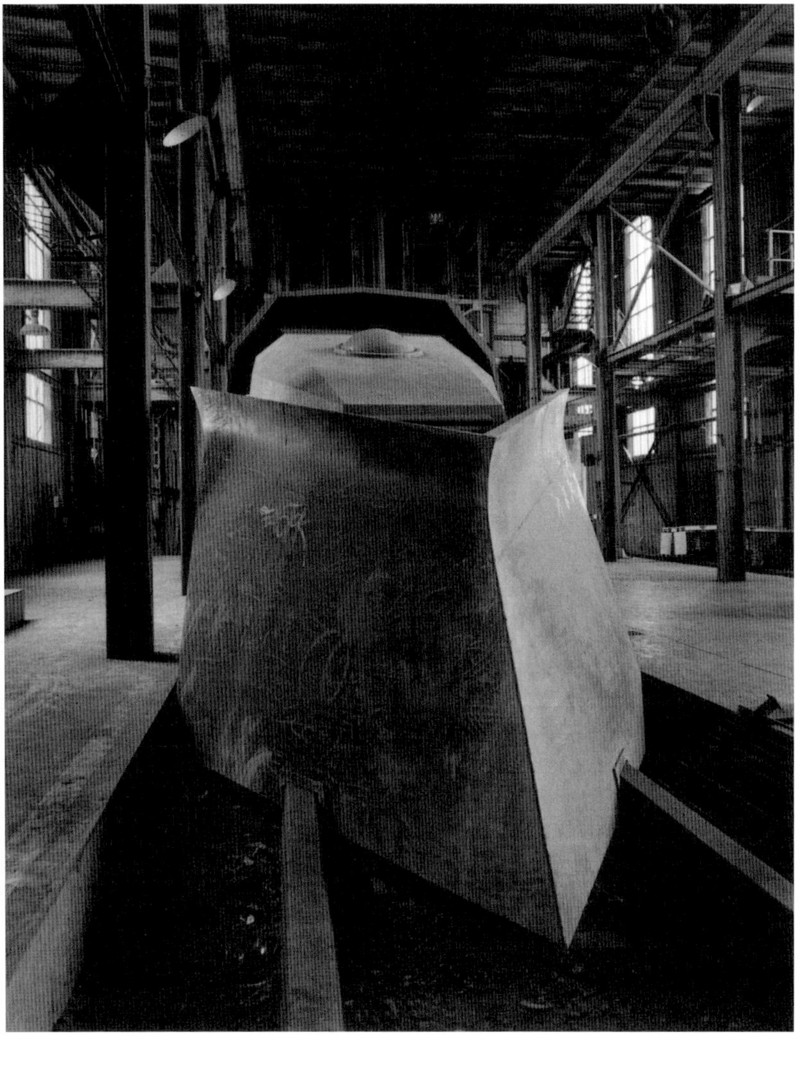

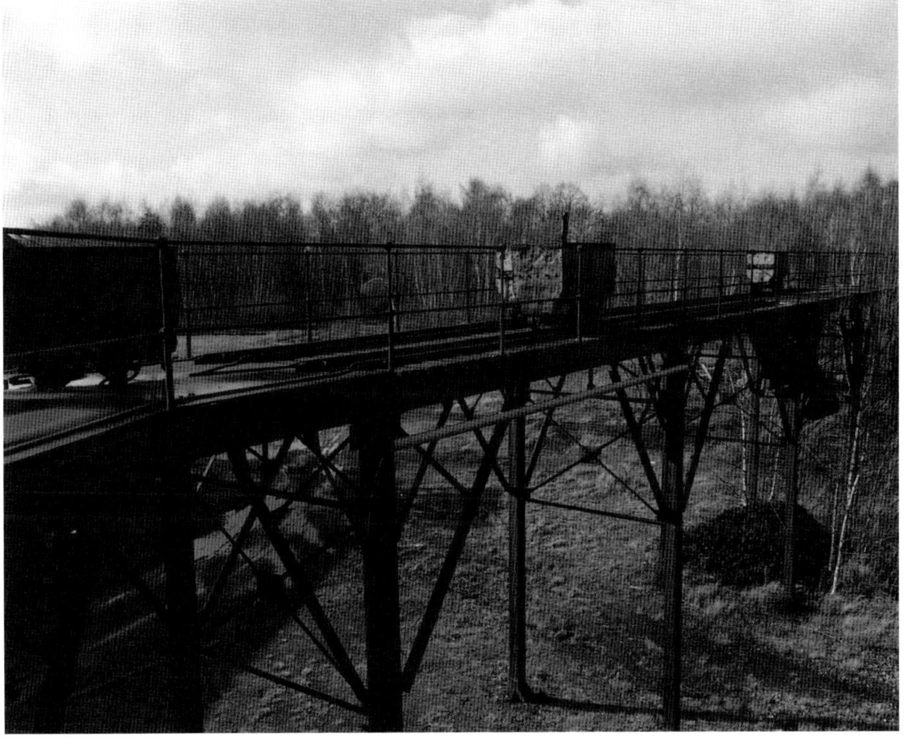

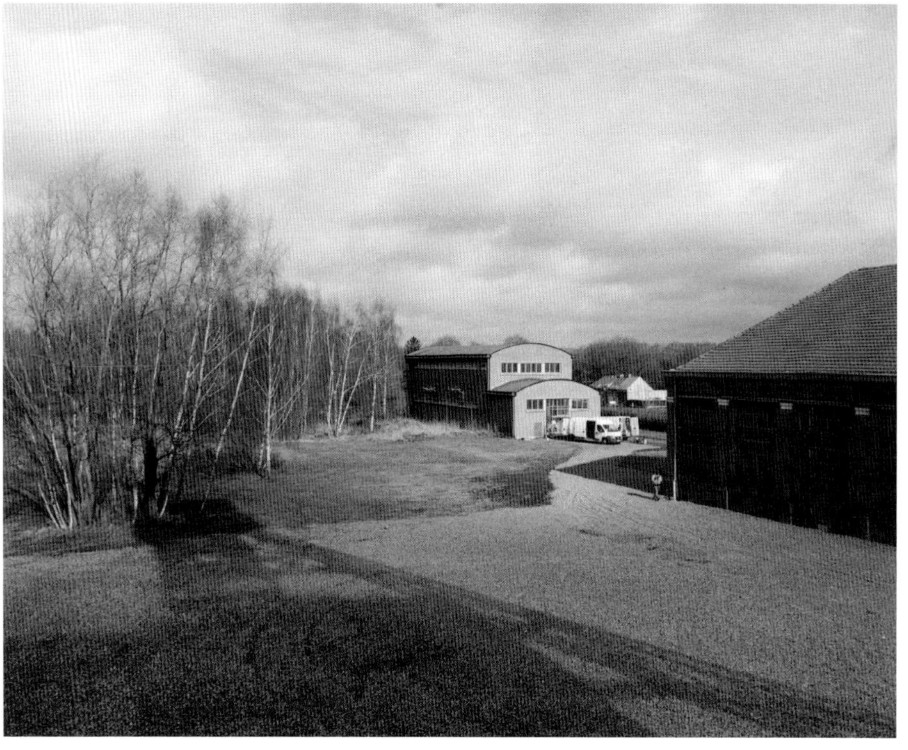

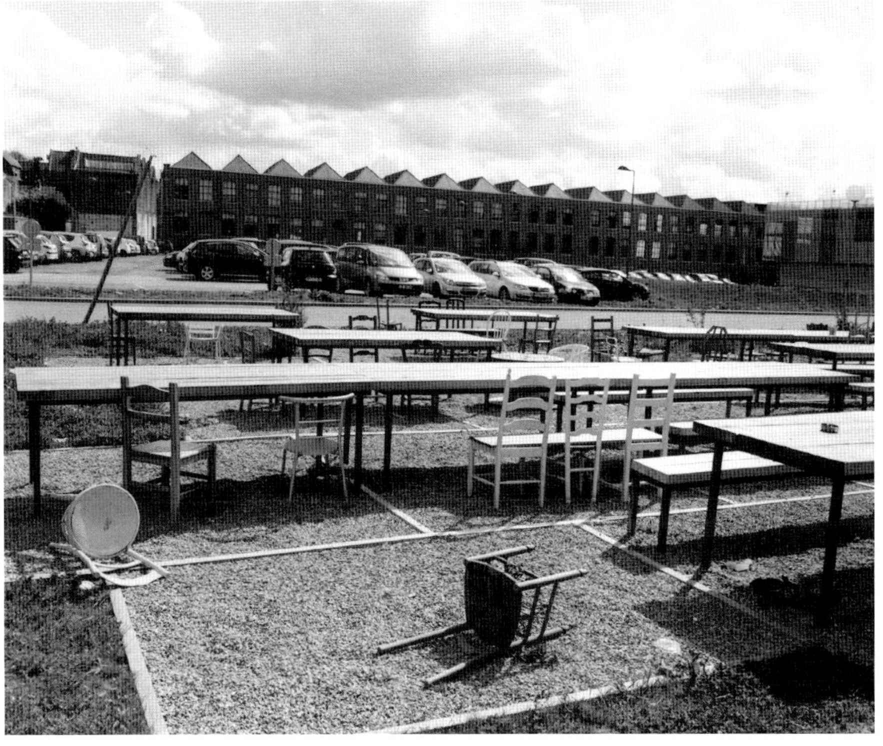

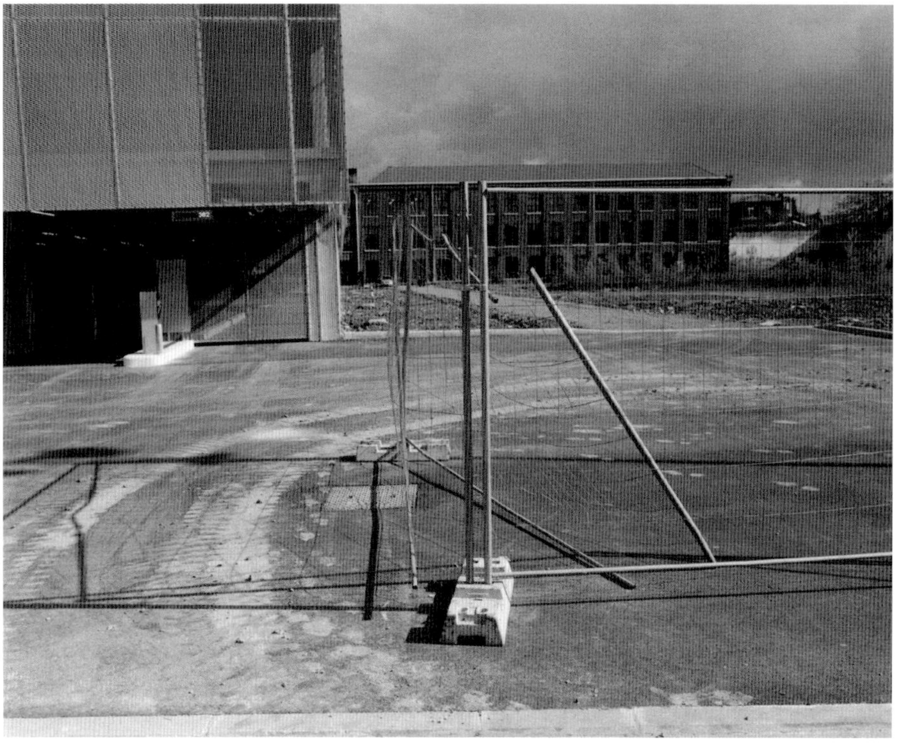

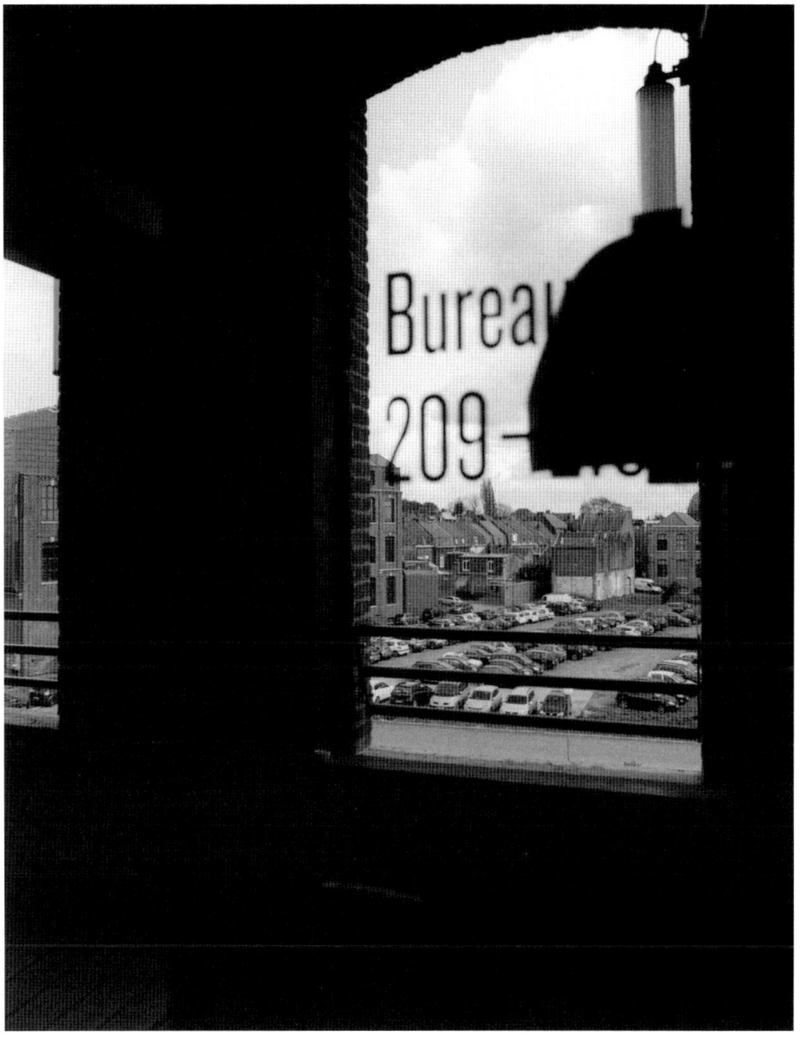
Bureau
209-

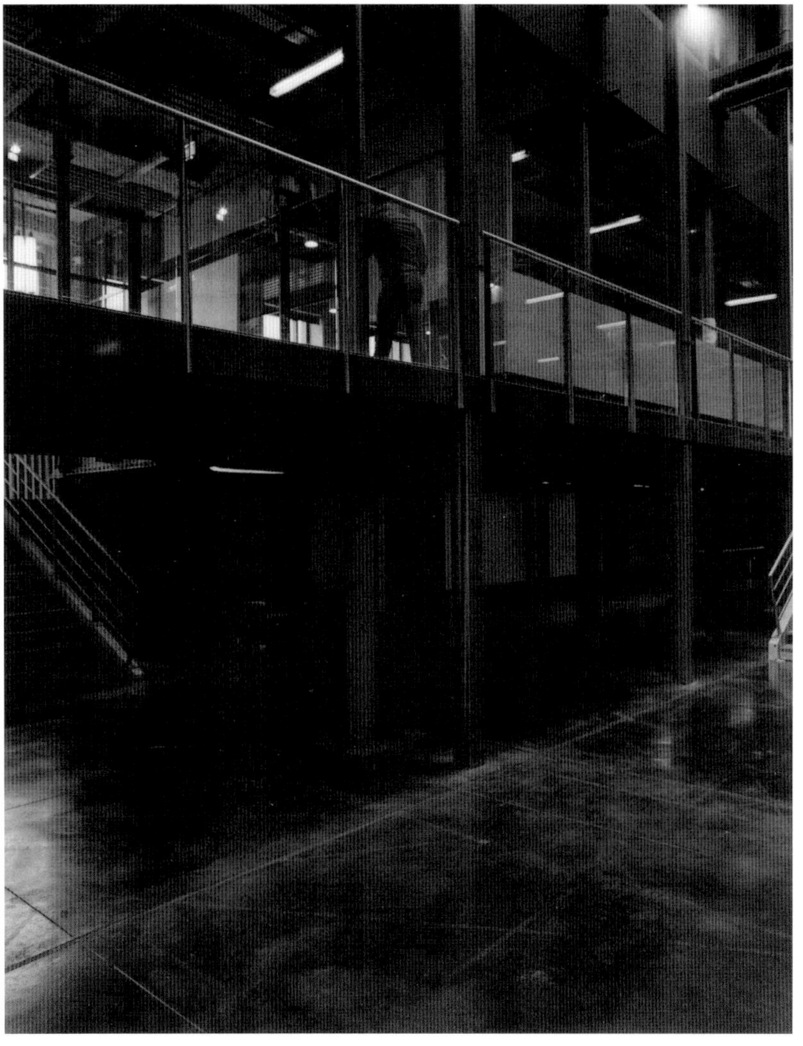

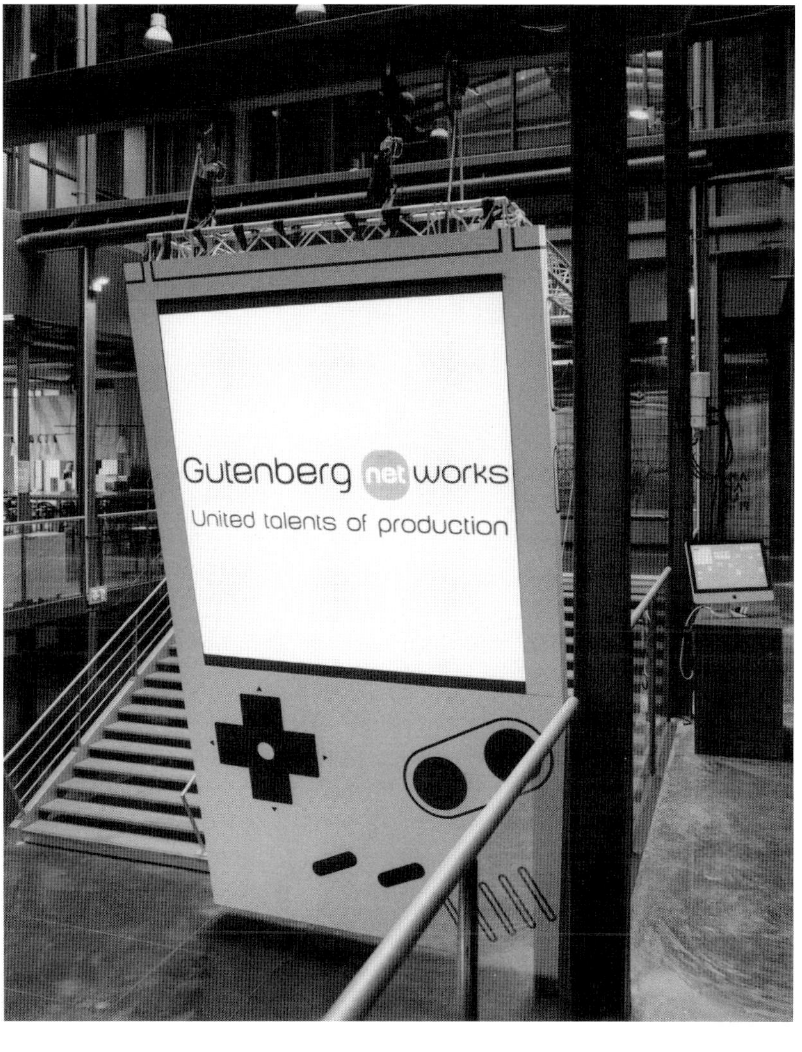
Gutenberg net works
United talents of production

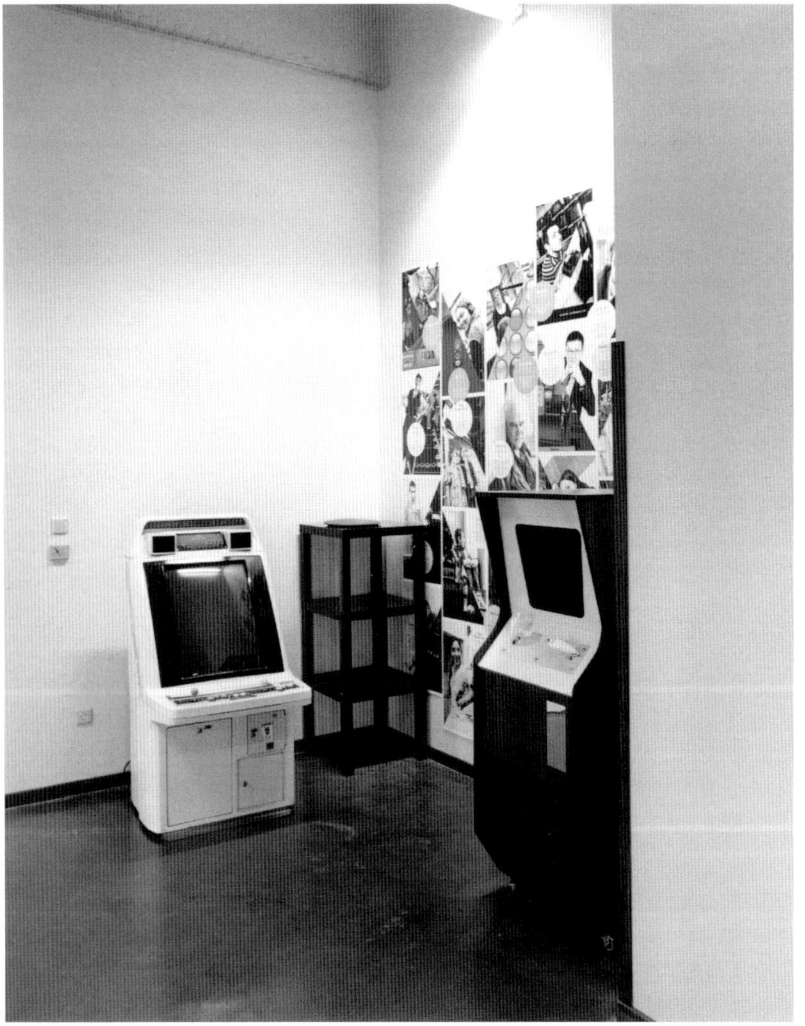

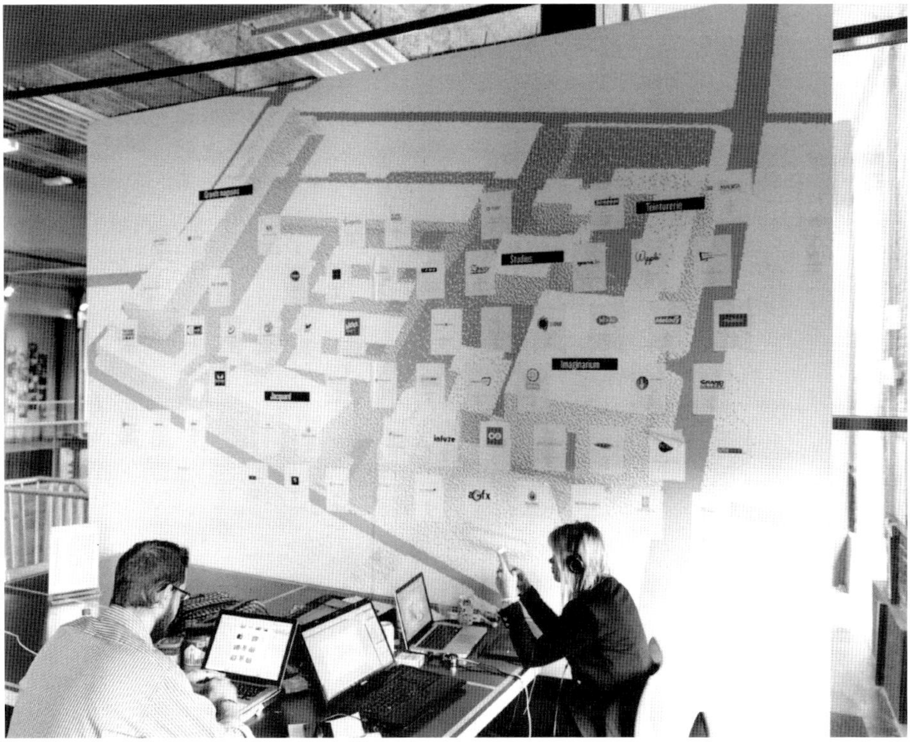
Teinturerie
Studios
Imaginarium
infuze

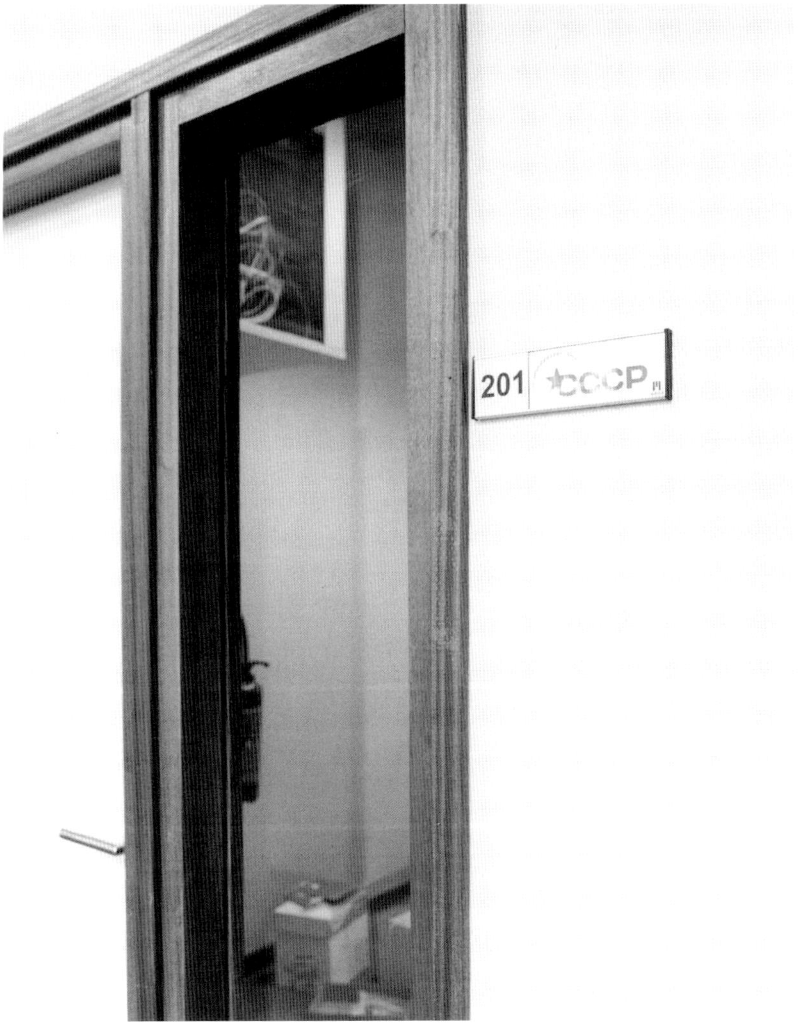
201
CCCP

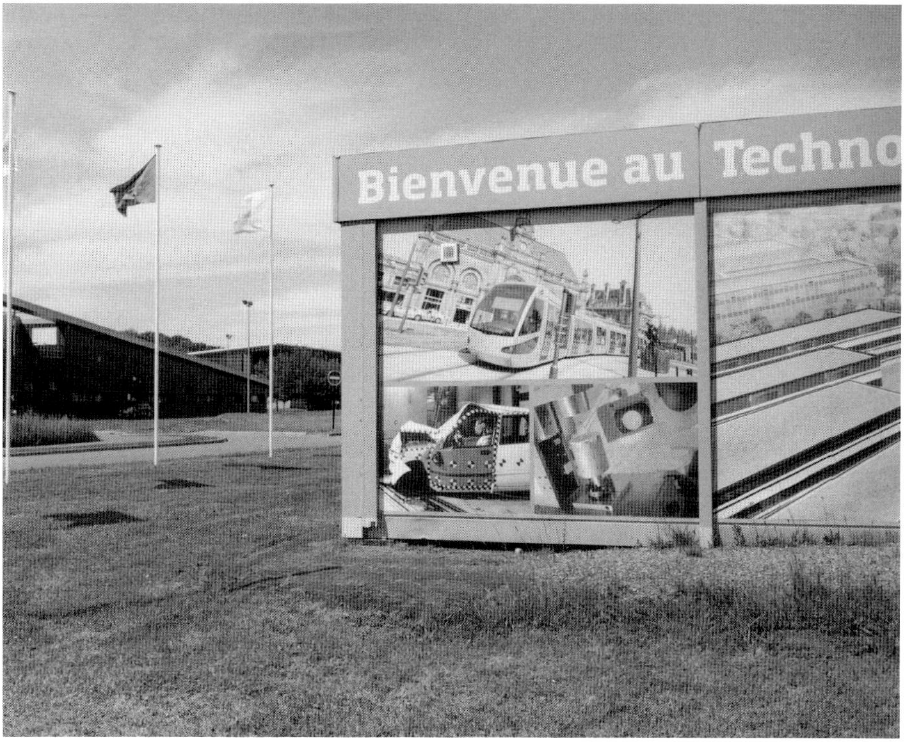
Bienvenue au Techno

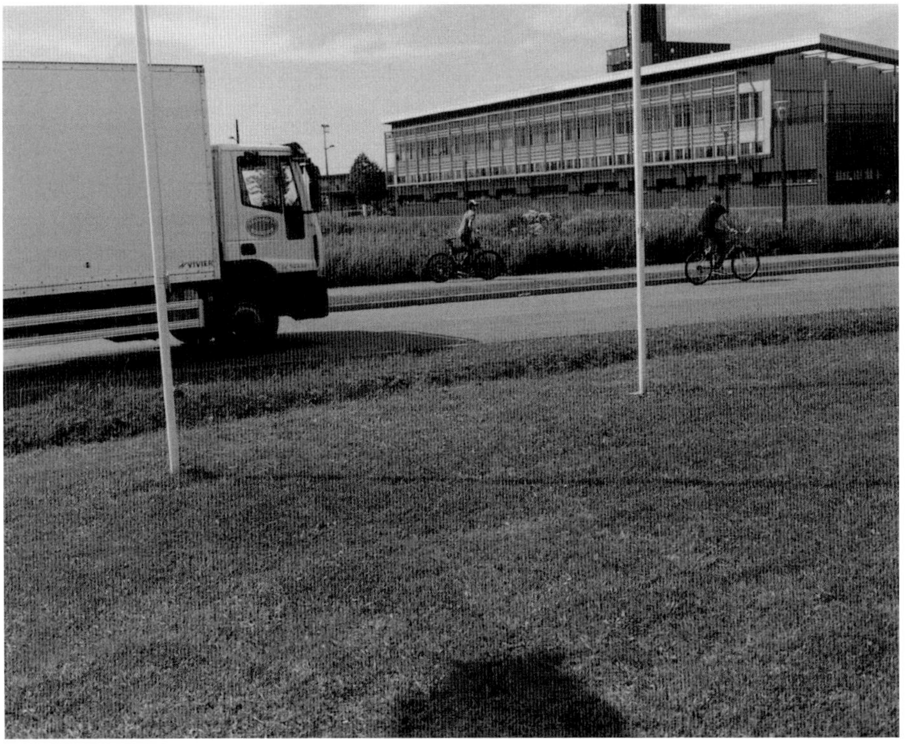

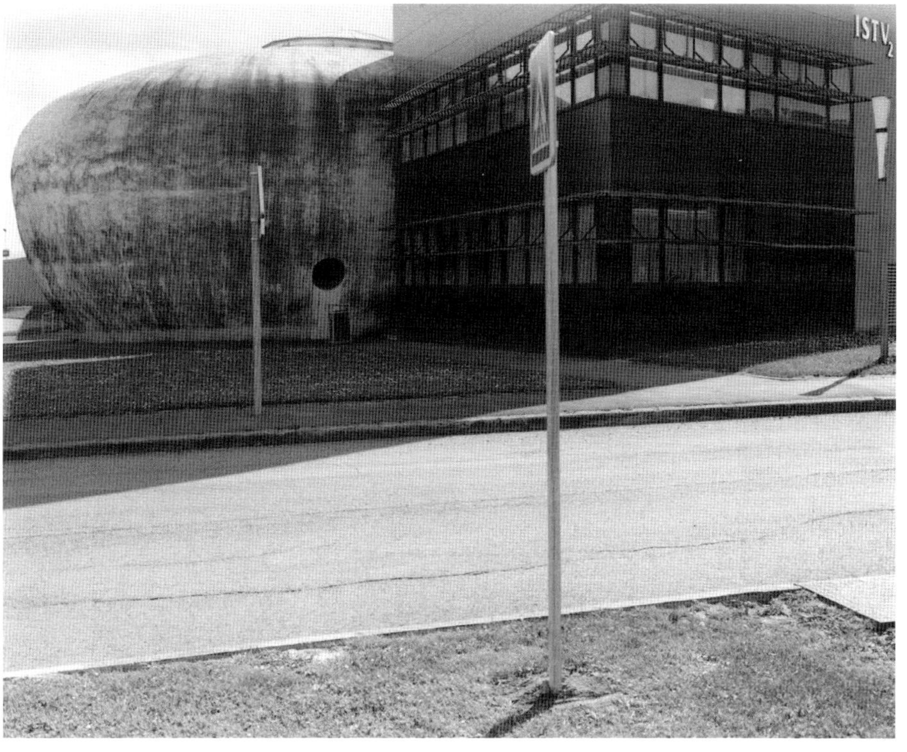
ISTV2

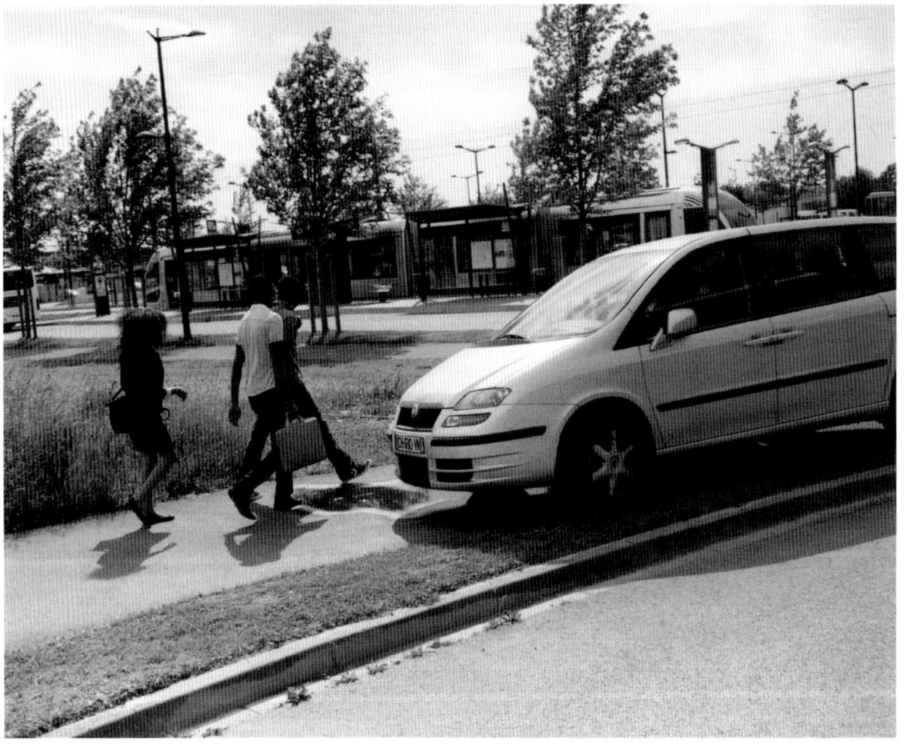

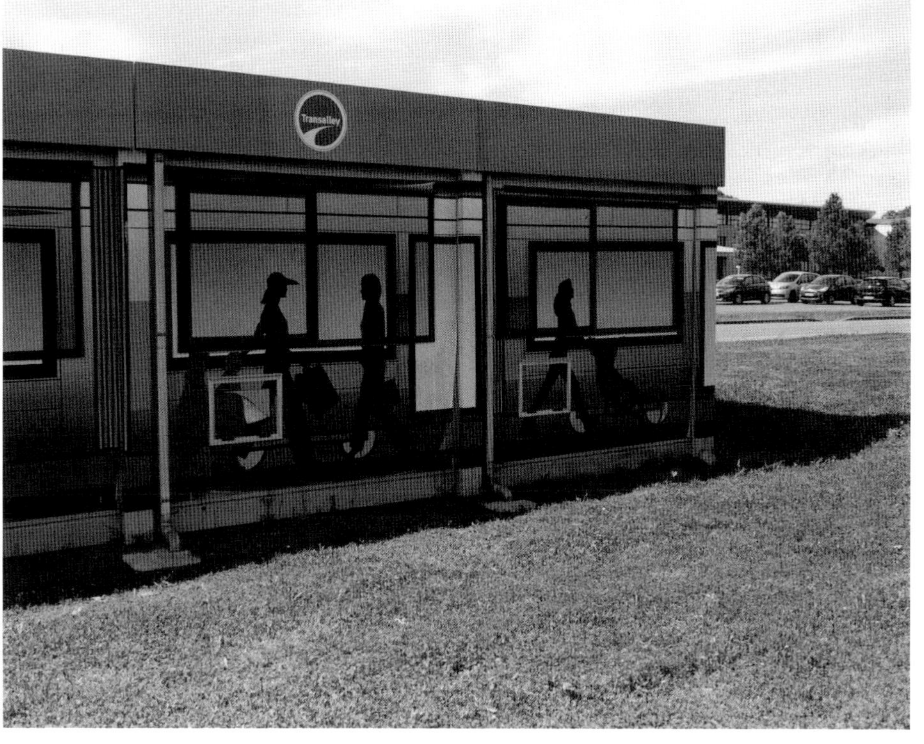
Transalley

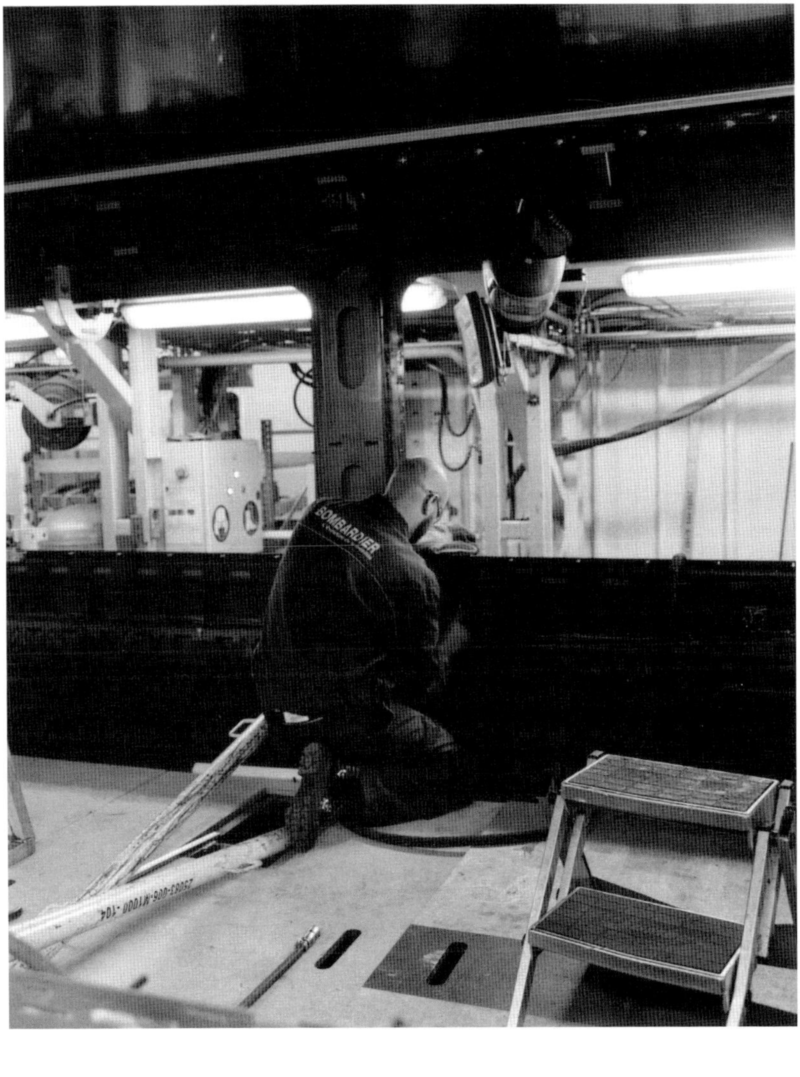
BOMBARDIER

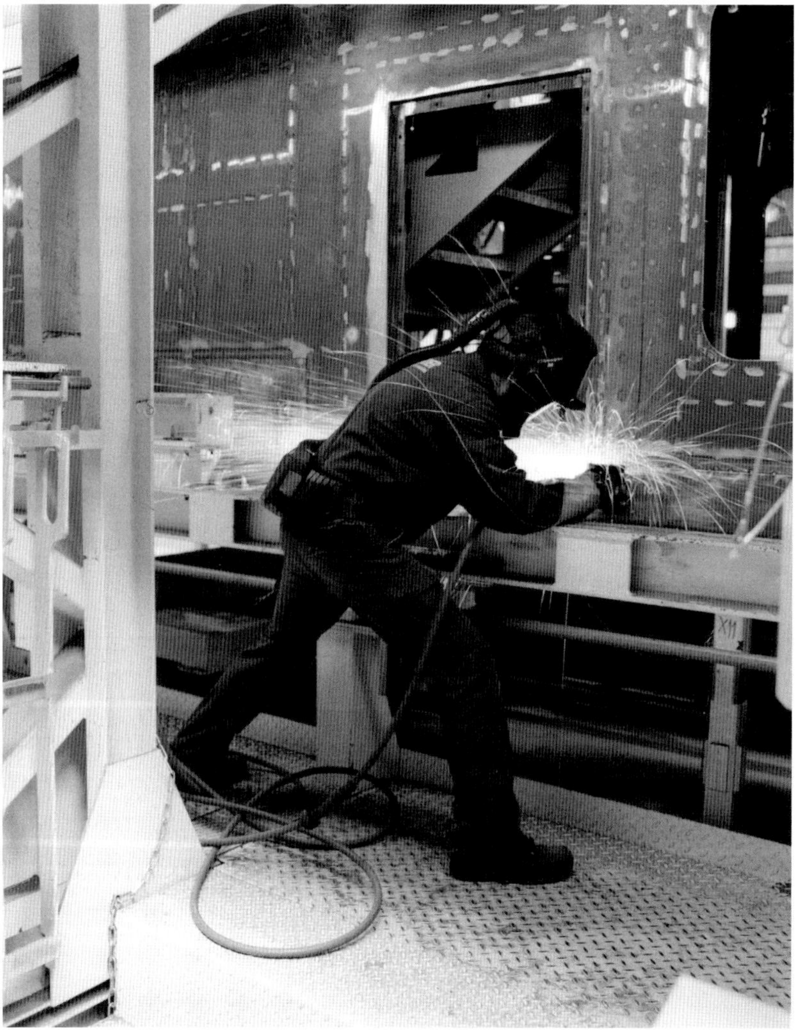

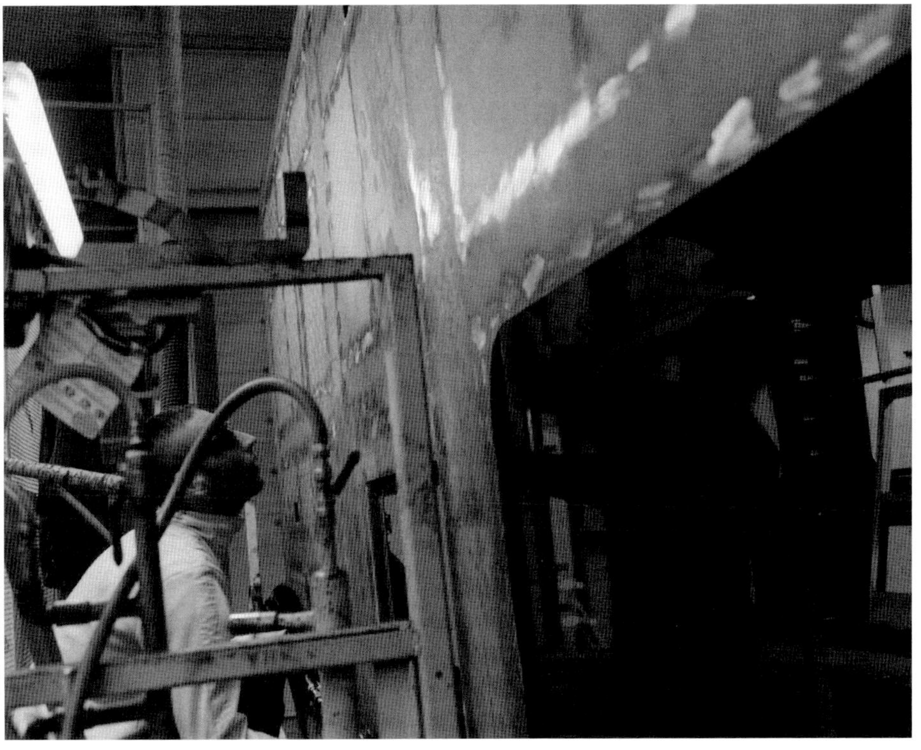

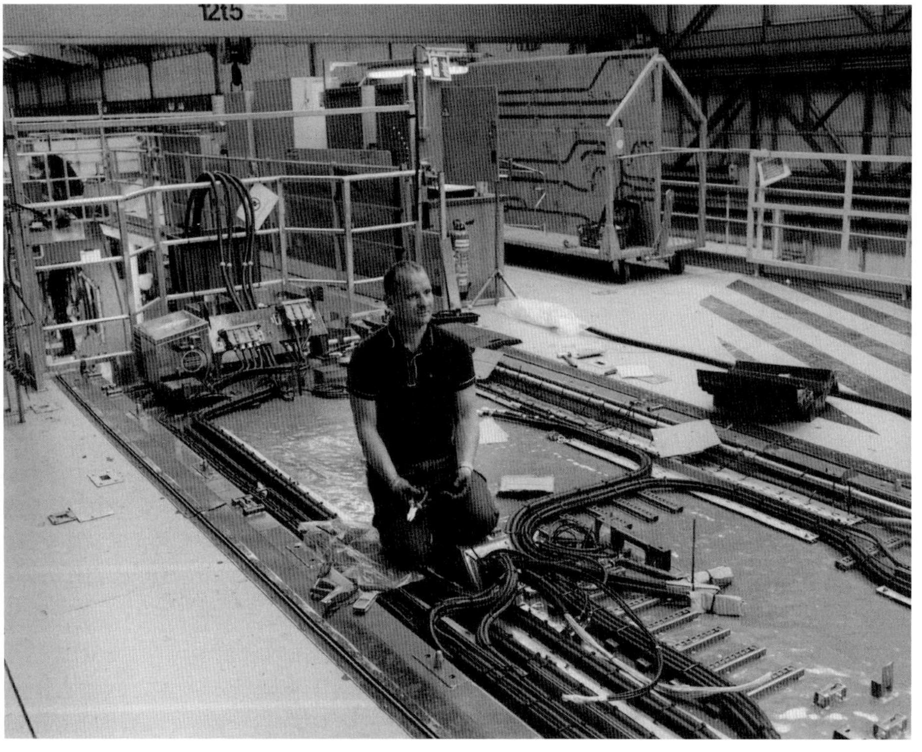
12t5

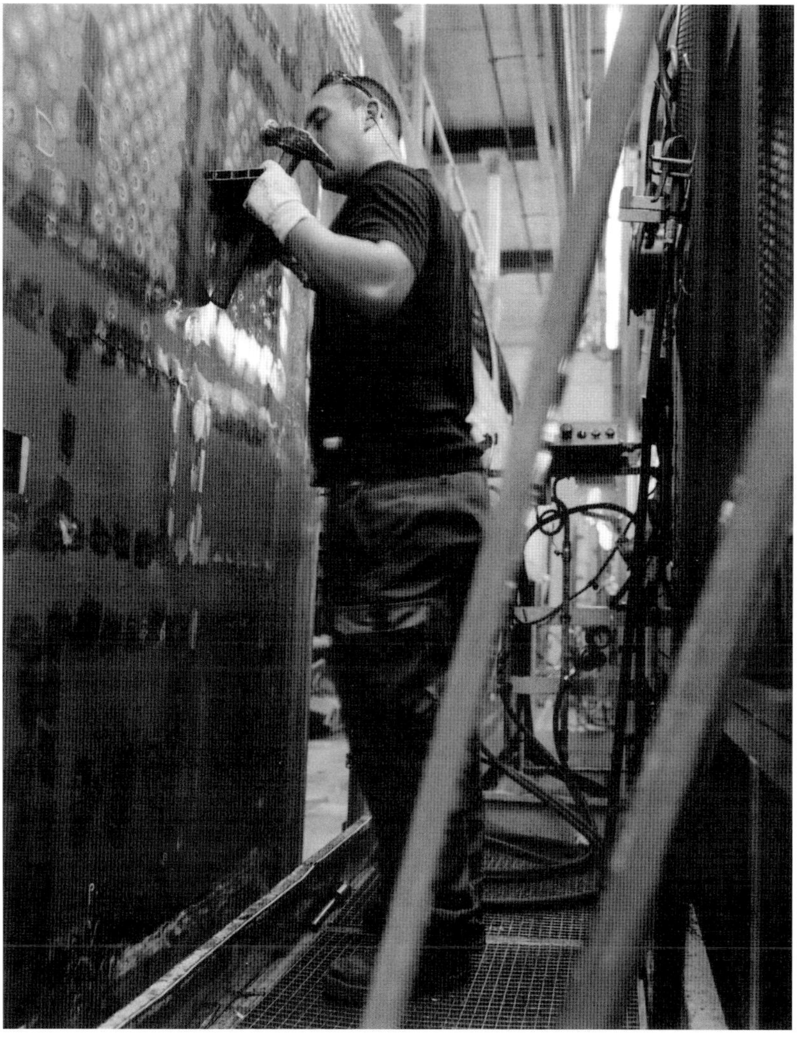

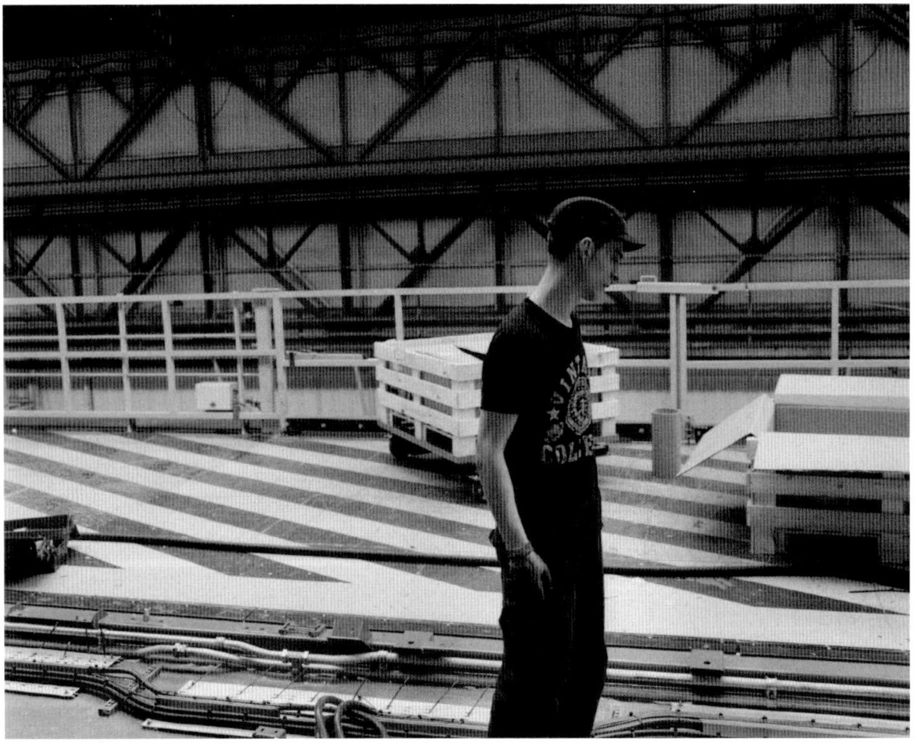

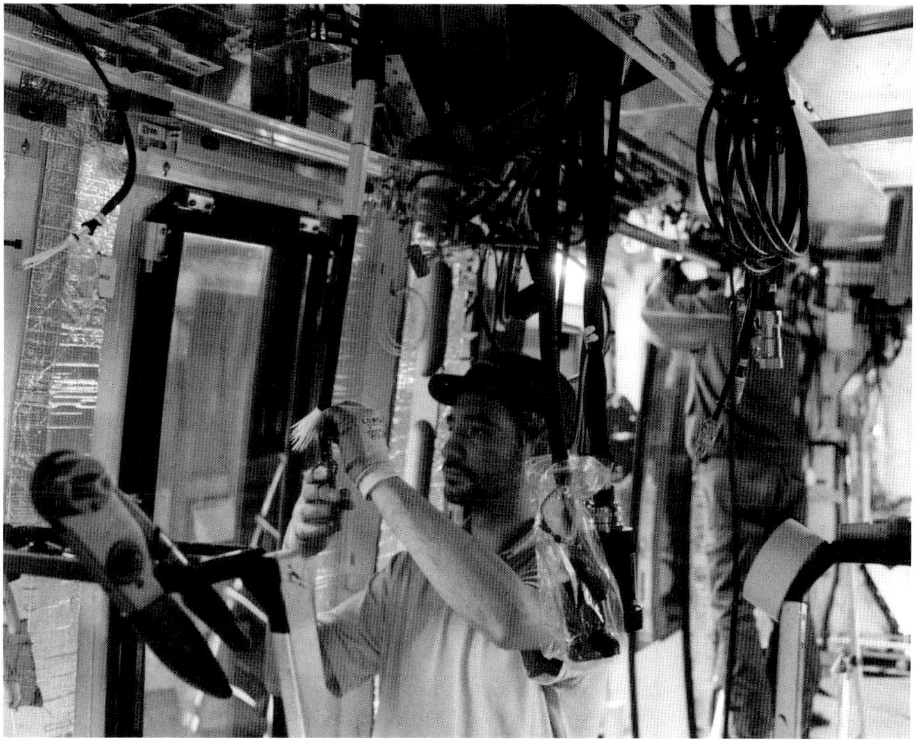

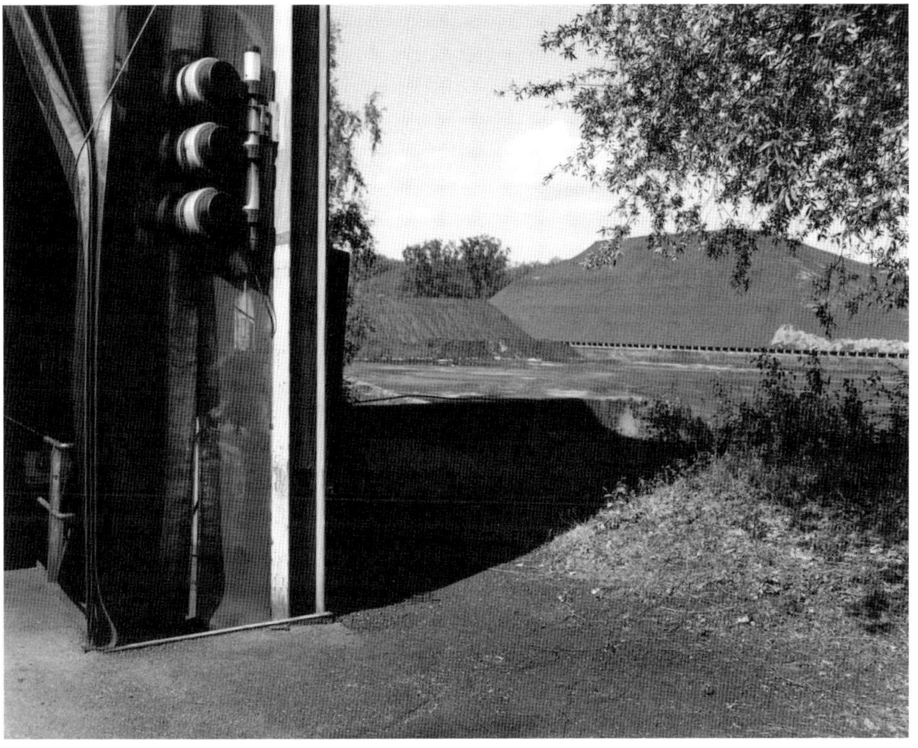

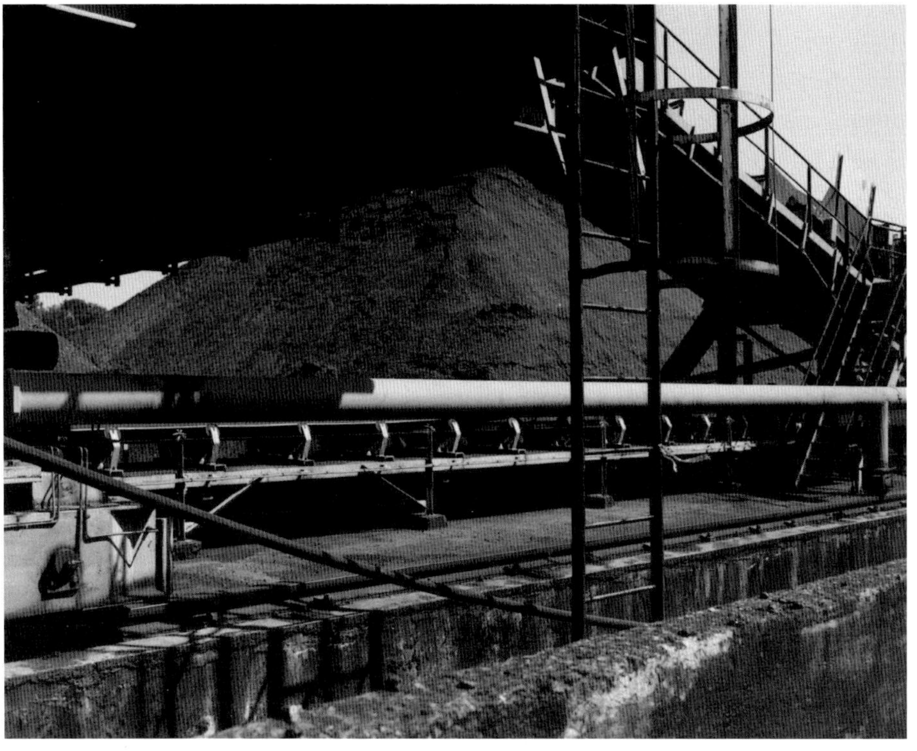

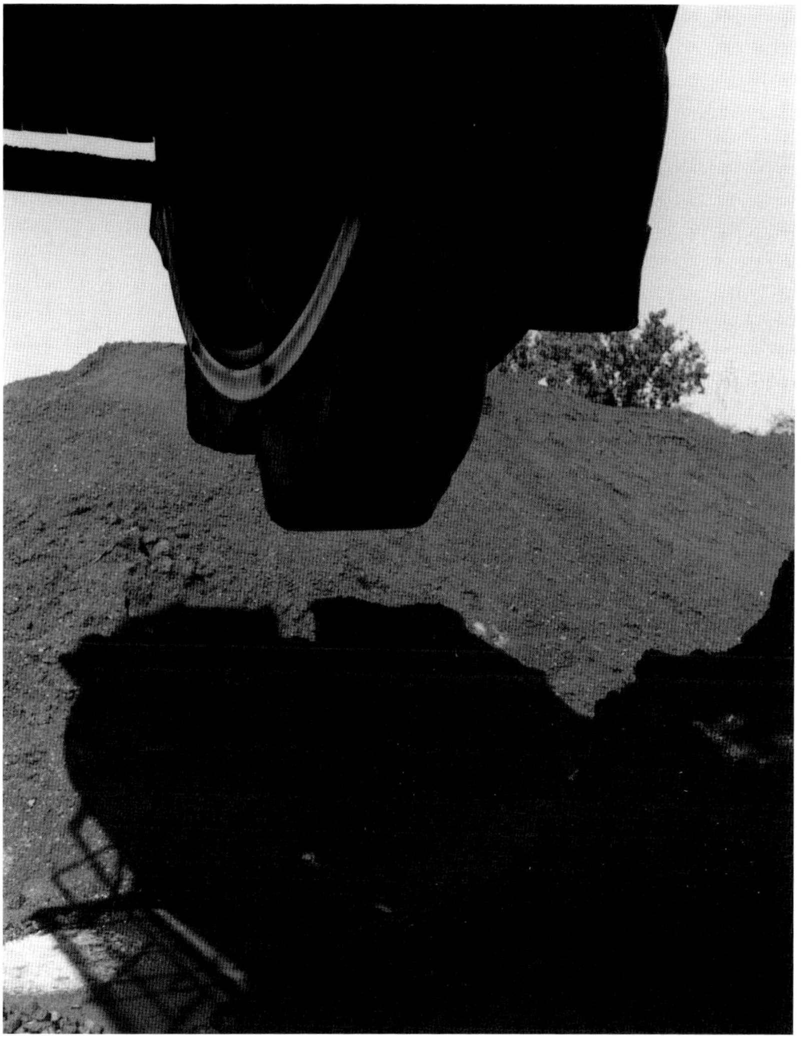

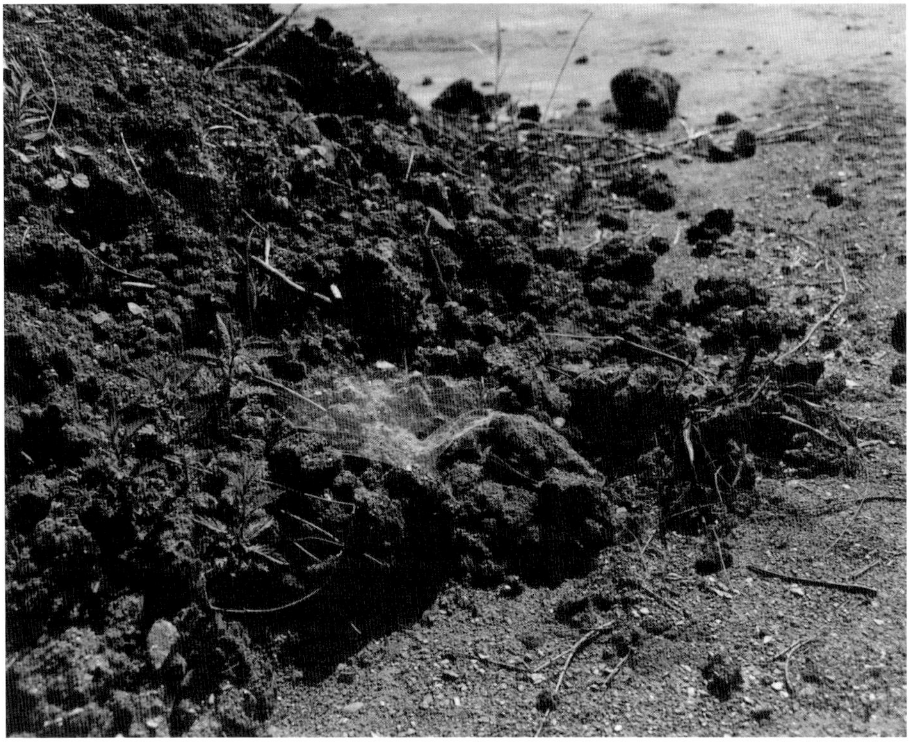

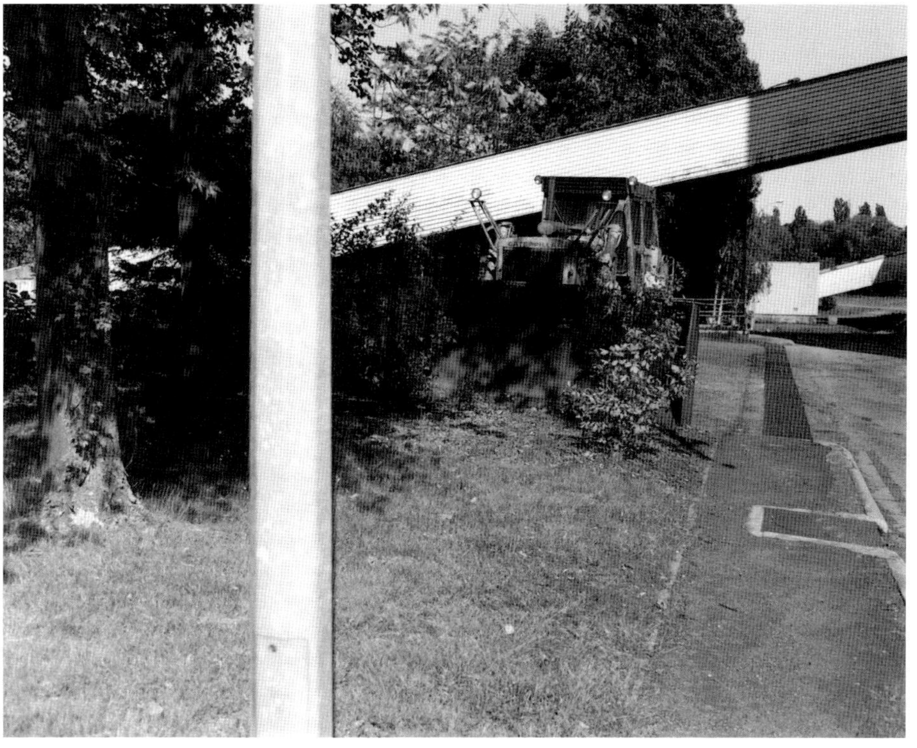

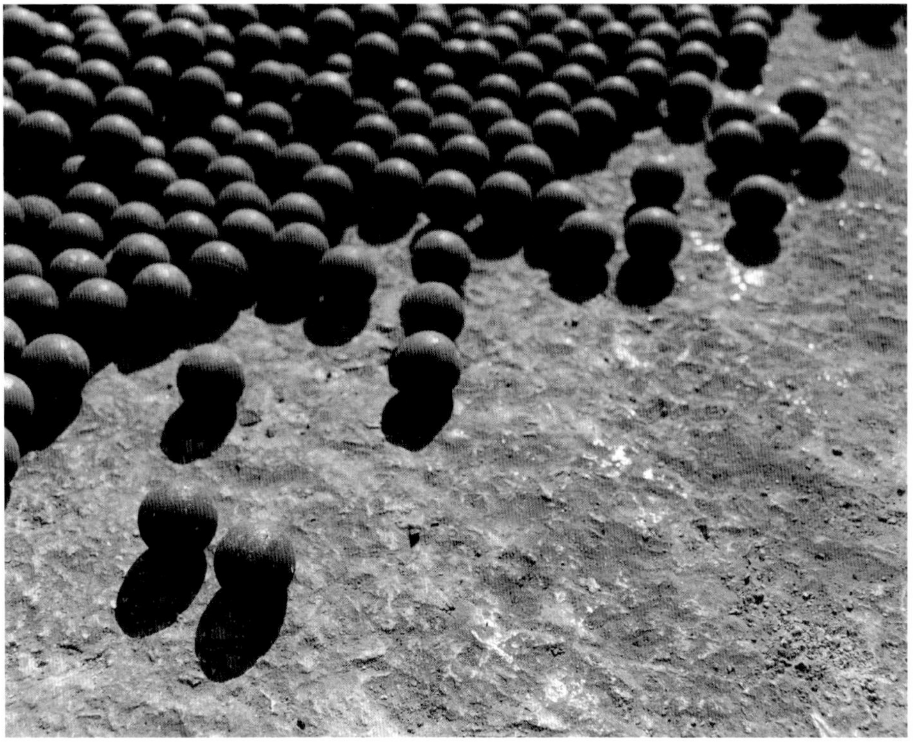

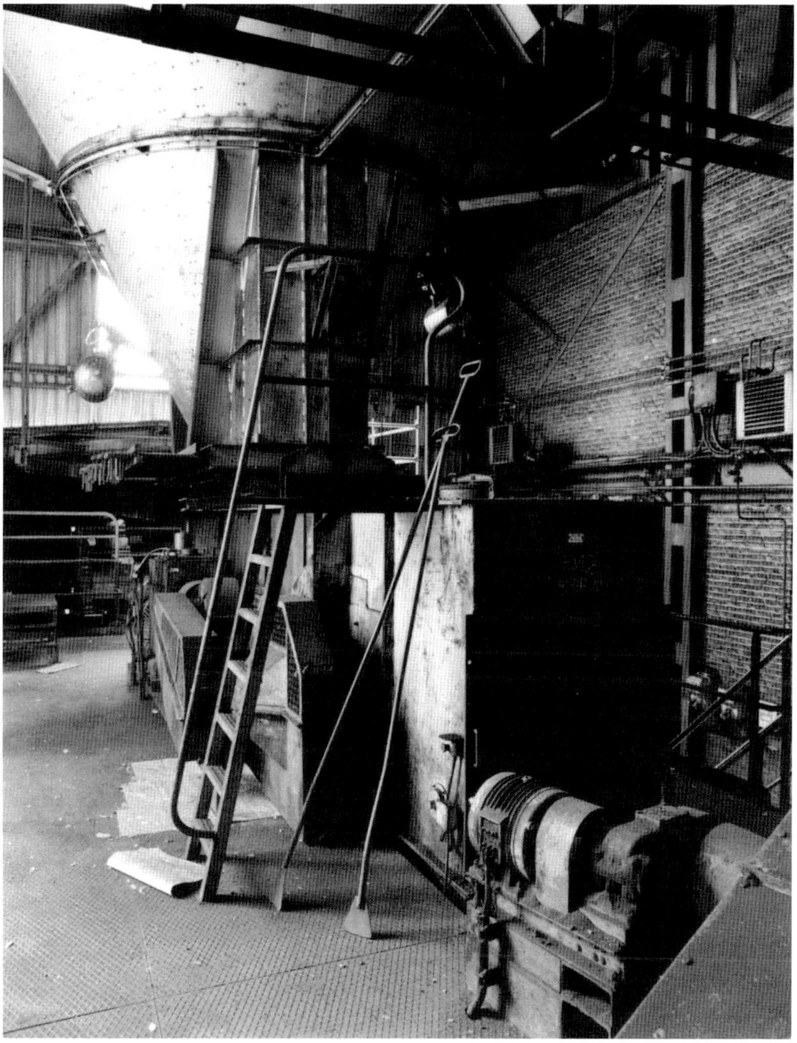

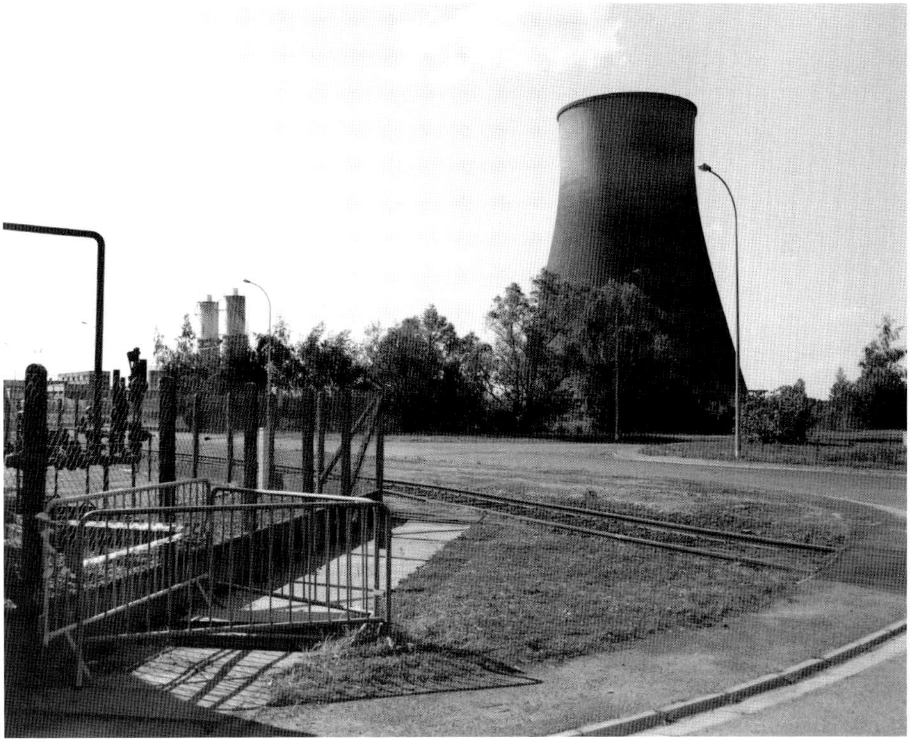

Leisure

The last scene is a tour of some landscapes of tourism and leisure related facilities in former mining sites. These are:

At the Lille airport arrivals hall, the advertising of the new Louvre-Lens welcomes the visitors. In the new Lille downtown, looking from the Pont de Flandres towards Euralille train station and shopping mall, an office worker rests at the lunch-time break.

In Nœux-les-Mines, Loisinord is a skiing resort built on a *terril*, one of those idiosyncratic artificial hills made of the mineral residues of the coal process that is part of the landscape of this and all coal mining areas.

Le Métaphone is an experimental 1000 seats popular music concert hall built in Fosse 9–9bis, Oignies, opened in June 2013. It is one of the most idiosyncratic new buildings in the site, conceived as an instrument itself: the building is covered by a skin of recycled objects and materials, including various instruments and a sound system. This way the building, designed by the Hérault-Arnod architecture office in collaboration with sound designer Louis Dandrel, can be literally "played".

On December 27, 1974 there was a mining accident at the Fosse 3 bis in Liévin. With 42 dead miners it was the biggest mining catastrophe in the region after the 1906 Courrières one. After an investigation and a trial, the managing enterprise Société des Houillères du Nord et du Pas-de-Calais was accused of negligence in 1981 and forced to cease activity. This was an important event for the ending of mining activity in the region. The tower has remained as a memorial of the catastrophe. Close to it there is a big shopping center with a Carrefour supermarket, built on the old site of the Fosse 1.

Close to the Liévin memorial and shopping center, in Lens, the newly built ultramodern Louvre-Lens, designed by SANAA architecture office, is surrounded by new landscaped gardens still in growth. The crude nakedness of the glass-iron minimalist design is reinforced by the recently planted, not yet blossomed gardening in the very early spring. The tour starts in the parking lot, continues in the outdoor cafeteria and the garden and plaza at the main museum entrance, and ends in the museum lobby, where an employee does cleaning work.

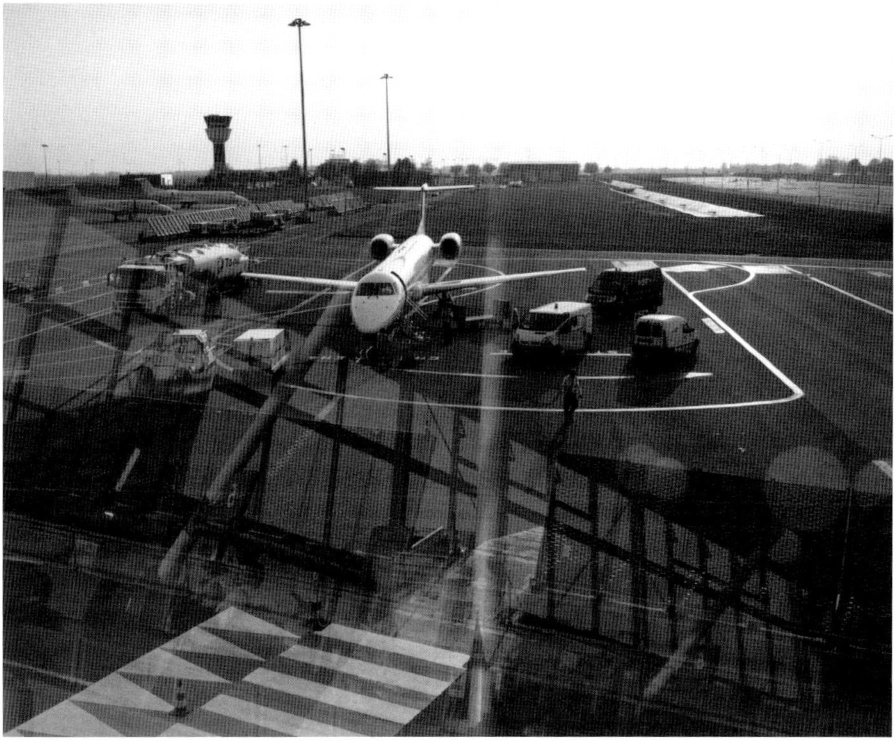

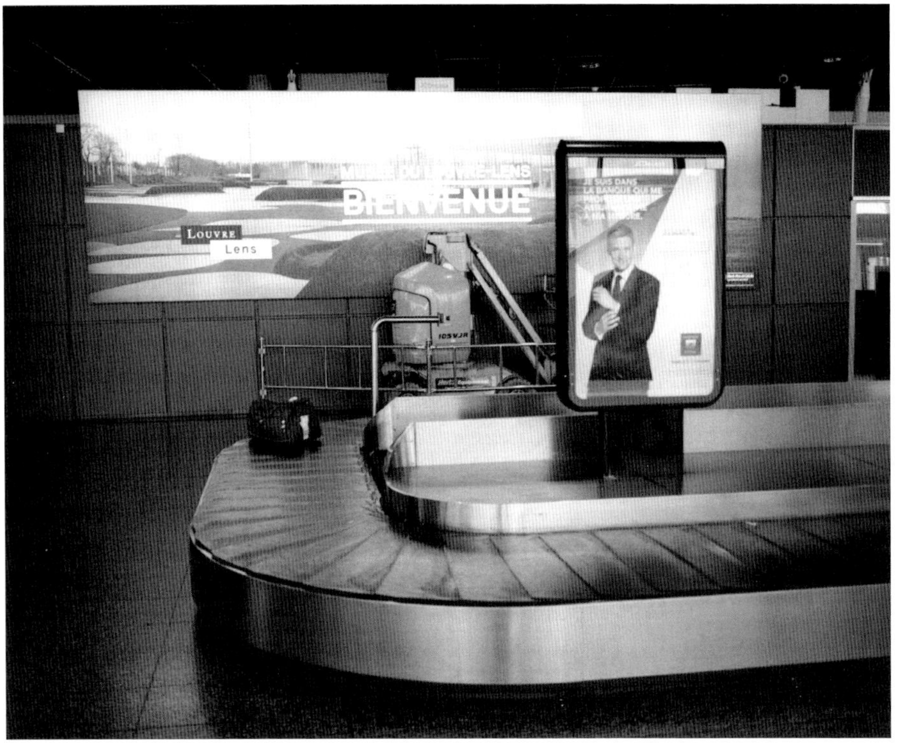
BIENVENUE
LOUVRE
Lens
105VJR

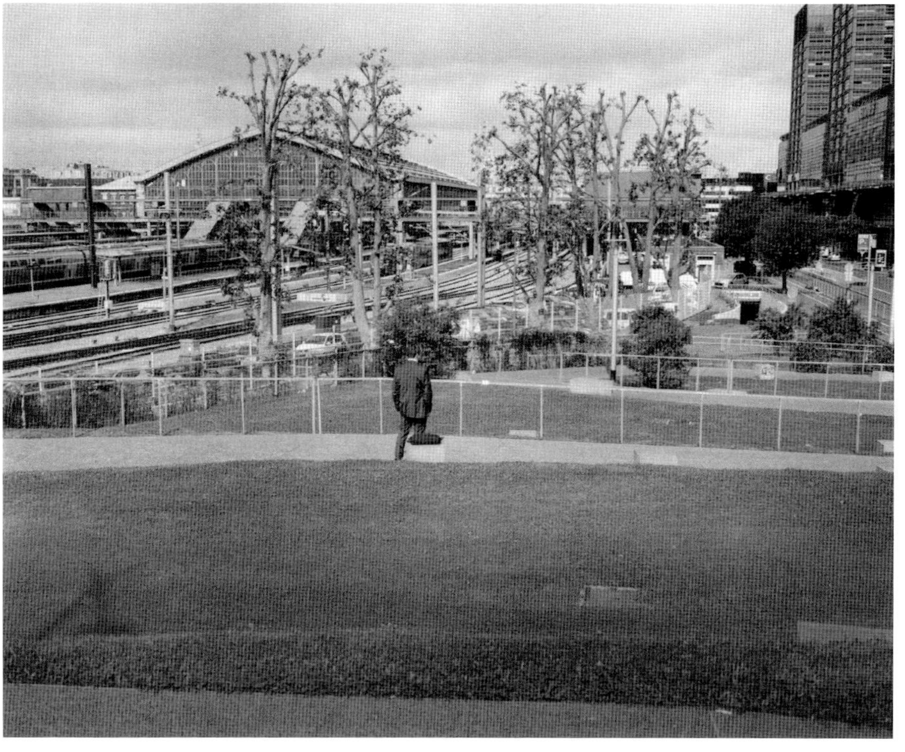

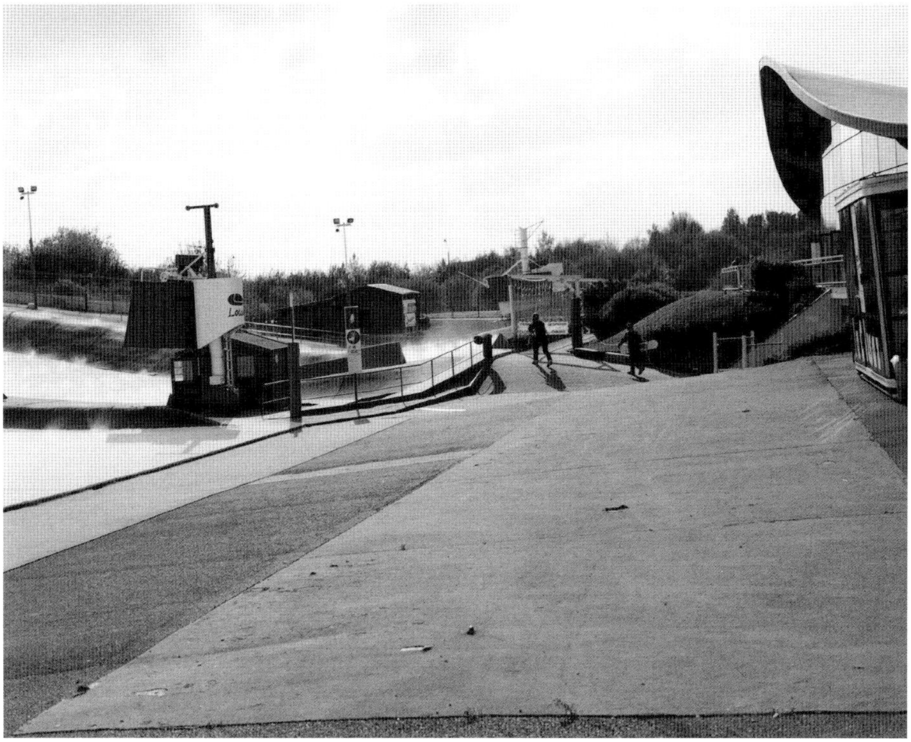

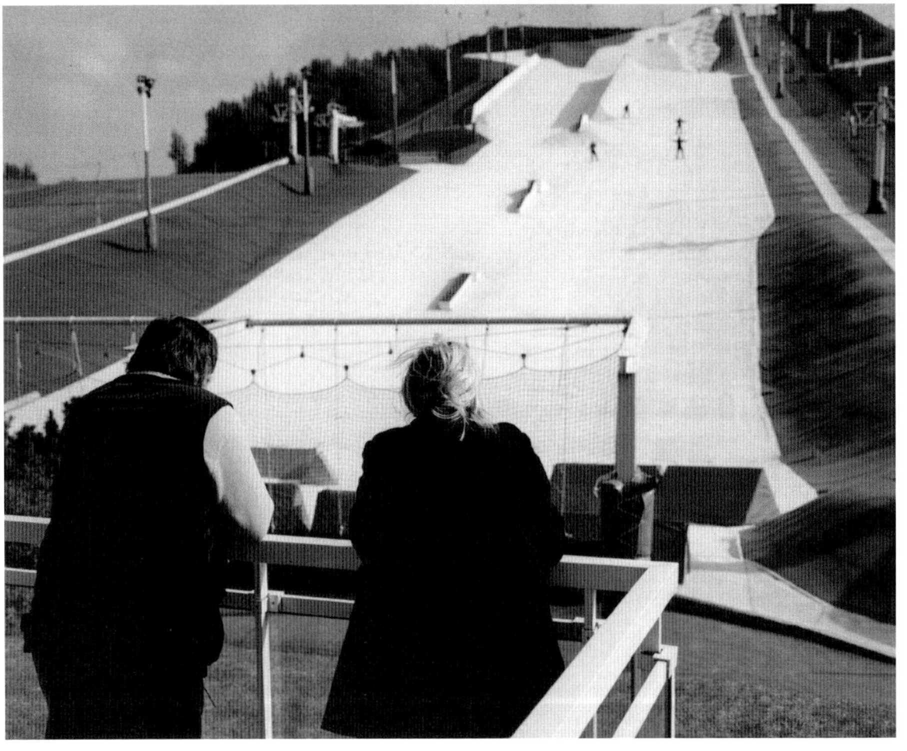

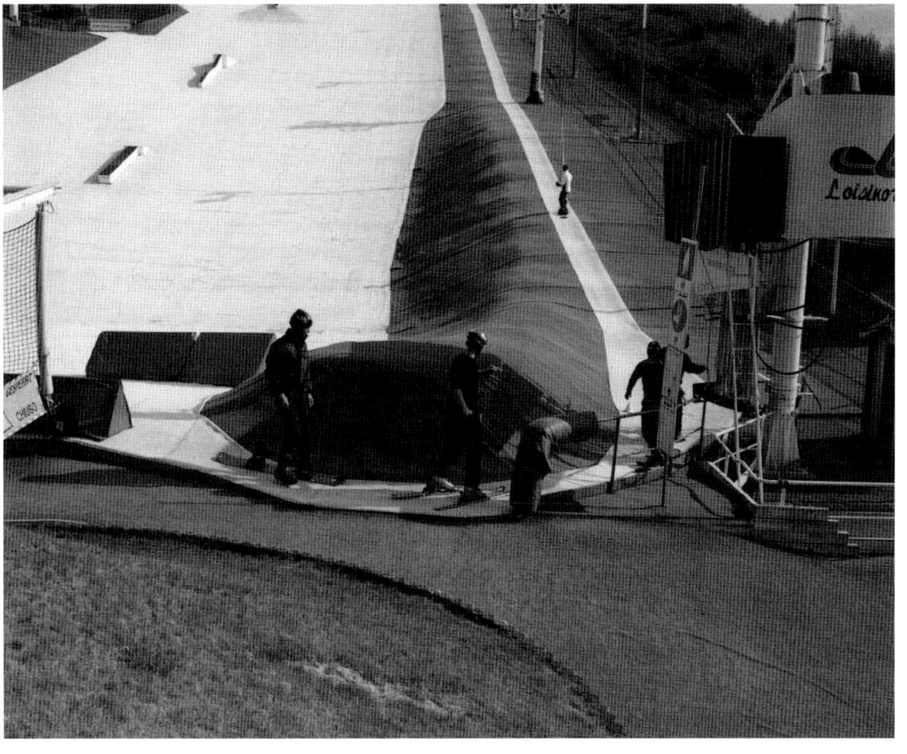

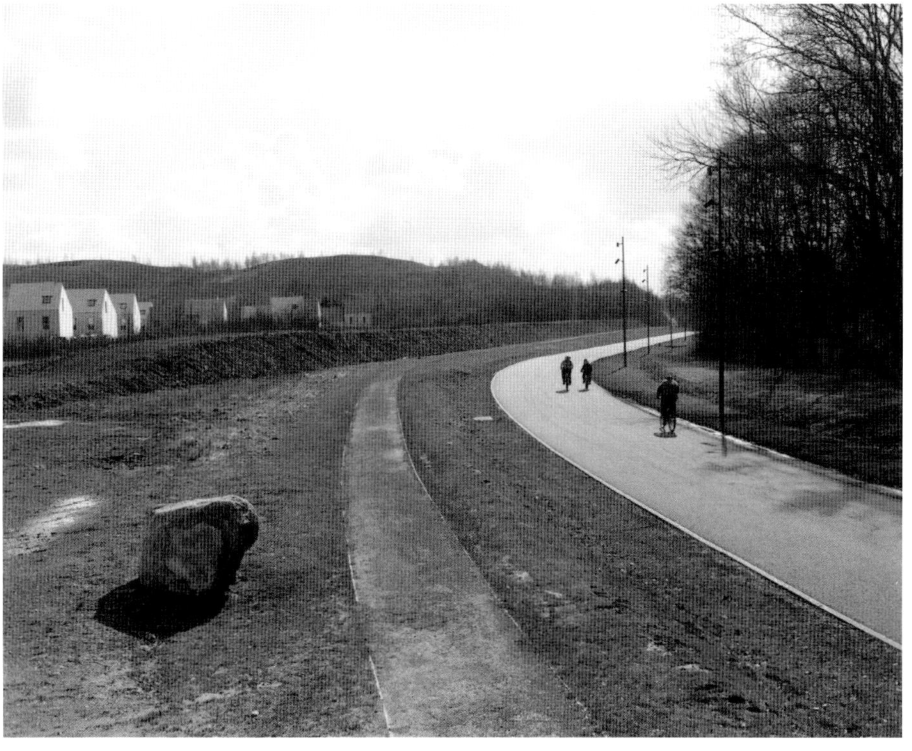

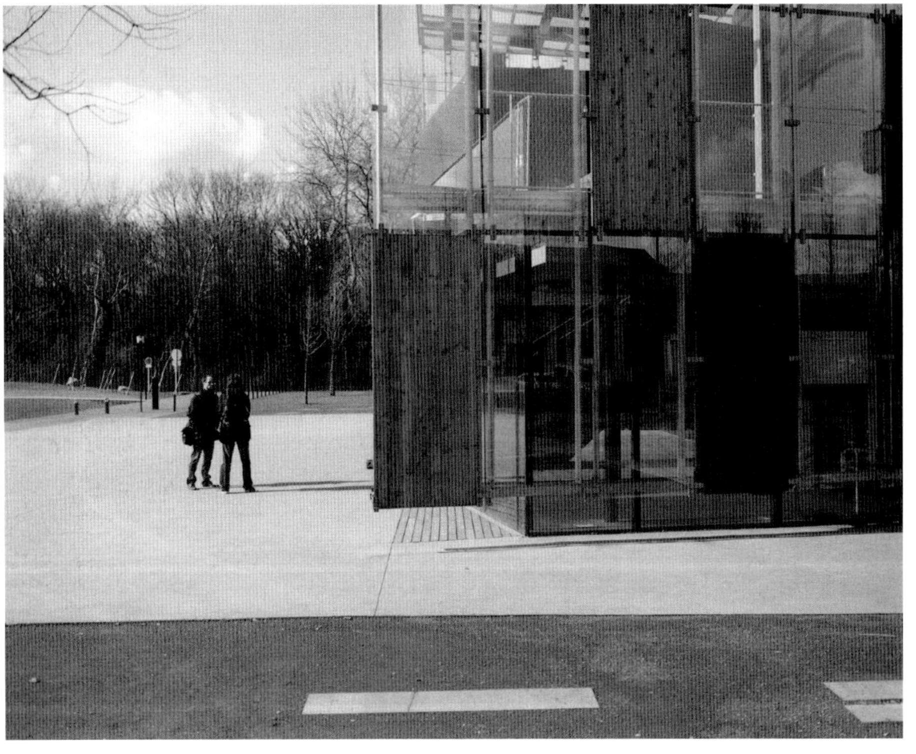

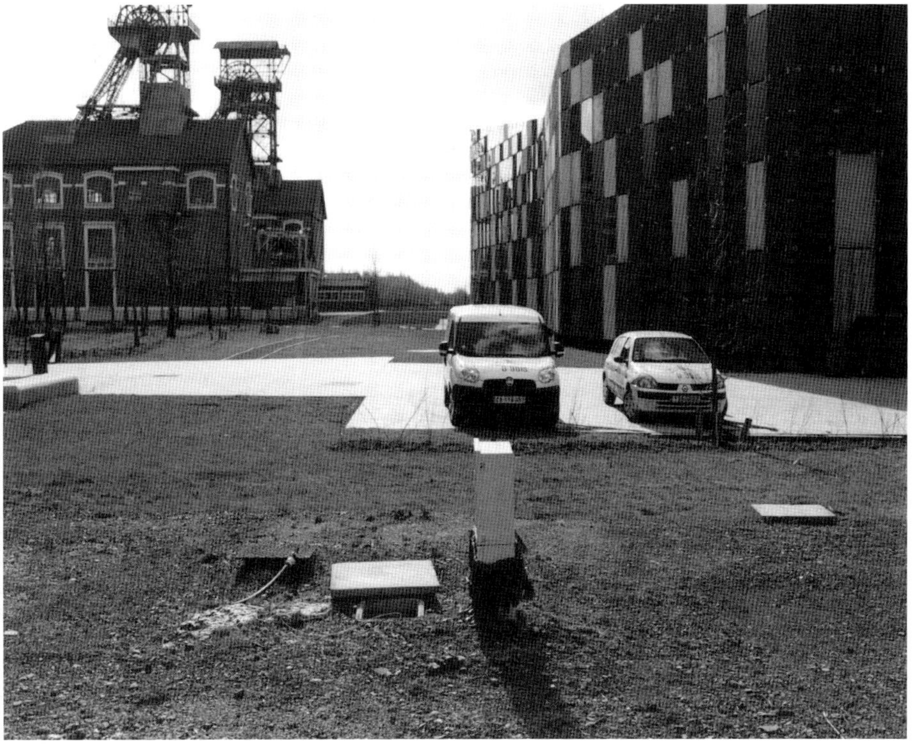

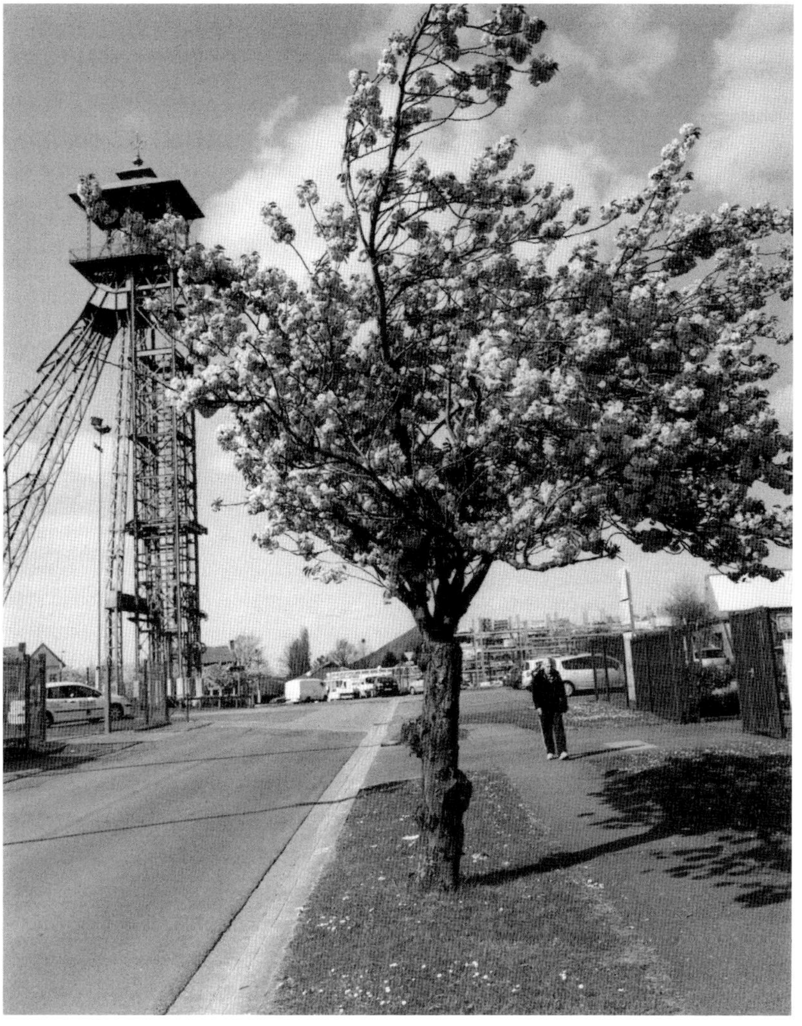

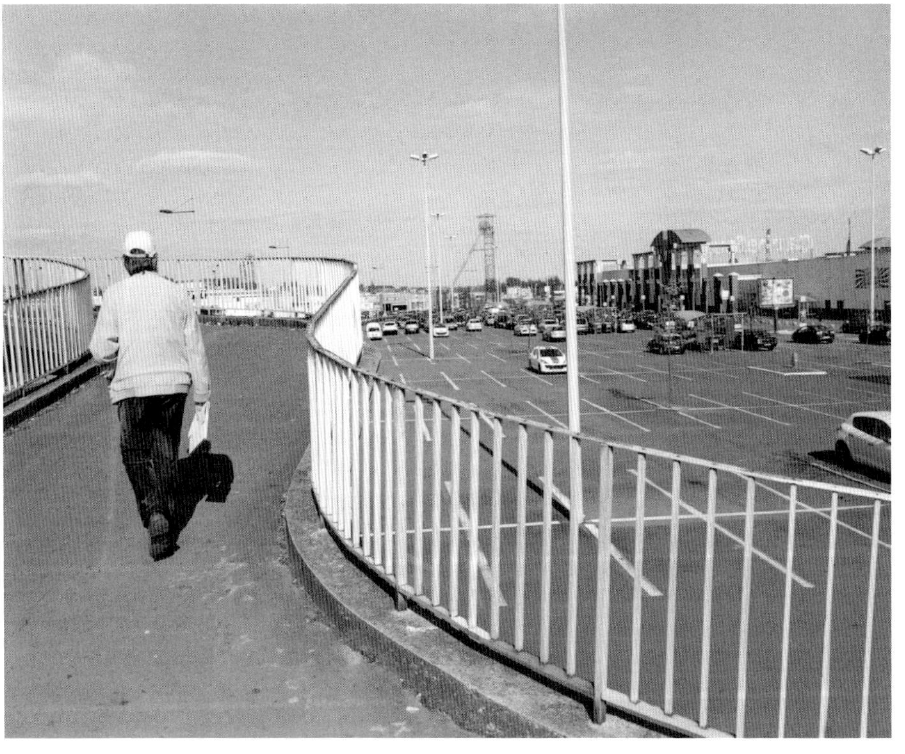

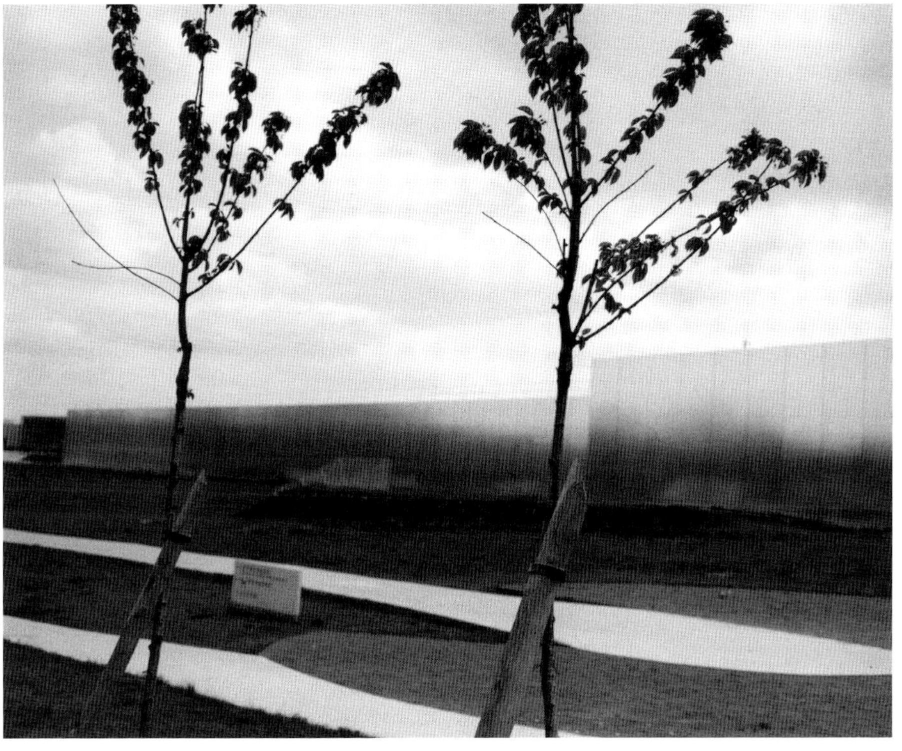

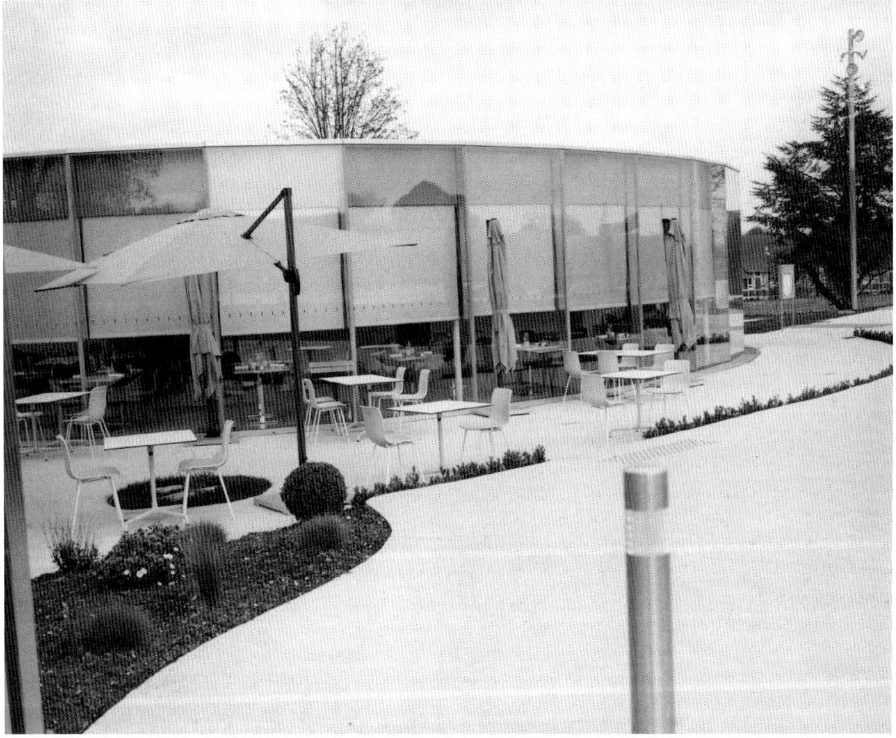

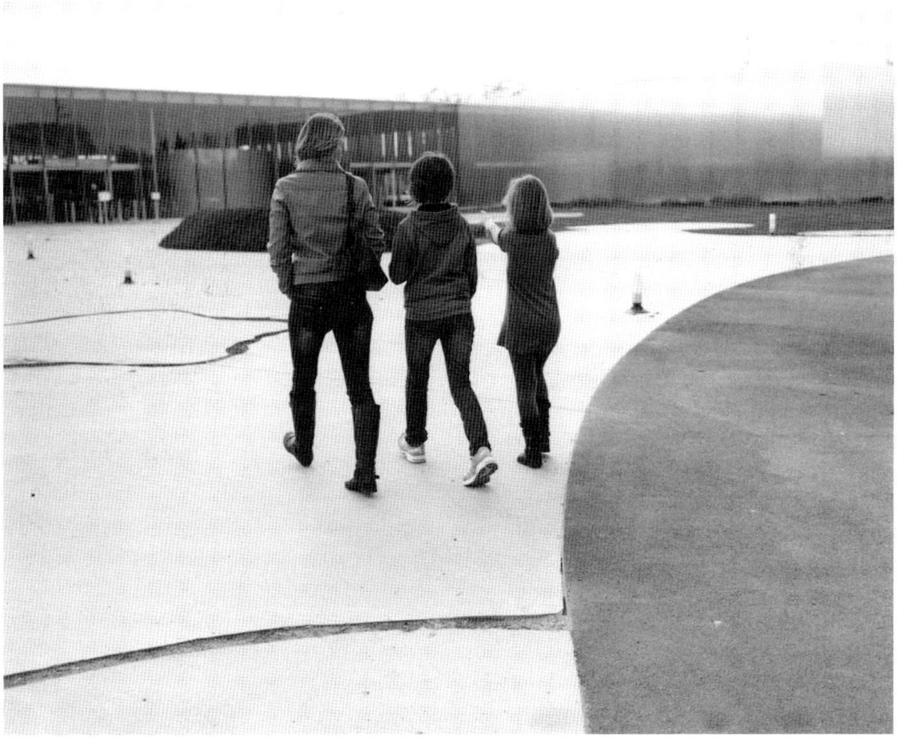

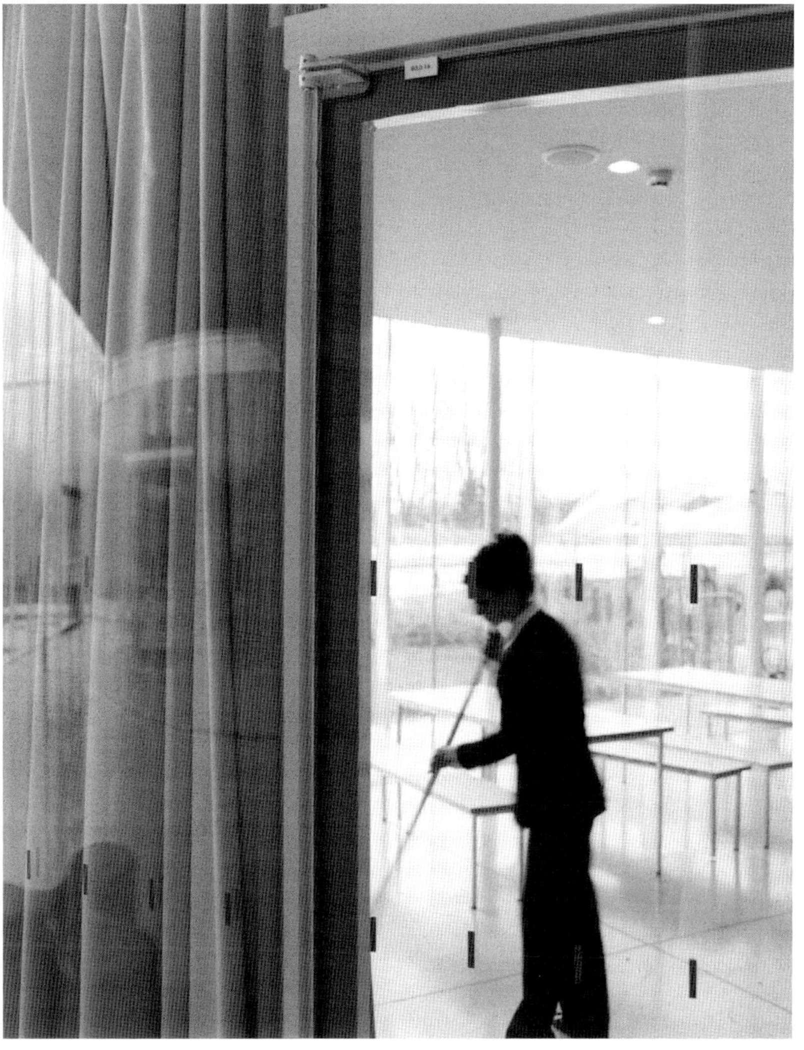

From Charles V to Louis XIV

13 Louvre-Lens, Renaissance gallery. Portrait of Anton Fugger by Hans Maler (c. 1529).

14 Maison Espagnole, tourist office, Valenciennes.

15 Porte Royale, Cambrai.

17 Henri Matisse, *Tête de faune* (1907), Musée Matisse, Palais Fénelon, Le Cateau-Cambrésis.

18 Museum shop, Musée Matisse, Palais Fénelon, Le Cateau-Cambrésis.

19 Garden of Palais Fénelon, Le Cateau-Cambrésis.

20 Noon, Musée Matisse, garden of Palais Fénelon, Le Cateau-Cambrésis.

22 Centre Historique Minier de Lewarde.

23 Entrance to exhibition *Le bassin minier, une conséquence de la bataille de Denain de 1712*, Centre Historique Minier de Lewarde.

25 Portrait of Louis XIV in the exhibition *Le bassin minier, une conséquence de la bataille de Denain de 1712*, Centre Historique Minier de Lewarde.

Small History of Photography

29 The album *La Belgique* (1854) by Blanquart-Évrard, open by Parc de Bruxelles plate, research facilities, Bibliothèque Municipale, Lille.

30 The album *La Belgique* (1854) de Blanquart-Évrard, open by Place Royale plate, research facilities, Bibliothèque Municipale, Lille.

31 *Album de l'artiste et l'amateur* (1851), first album by Blanquart-Évrard, photo archives, Musée d'Histoire Naturelle, Lille.

32 *Intervention de l'art dans la photographie* (1864), by Blanquart-Évrard, photo archives, Musée d'Histoire Naturelle, Lille.

34–35 2, rue Lombard, site of former Musée Industriel, Lille.

37–39 Surroundings of Canal de la Deûle, by avenue Kuhlman and route départementale 48, possible location of the former Blanquart-Évrard printing facilities, Loos.

41 Enlarger, former darkroom, CRP/, Douchy-les-Mines.

42 Pia Viewing, former CRP/ director holding a demolition shot from the exhibition *La mine en œuvre 1890–1990*, a collaboration of CRP/ with the Centre Historique Minier de Lewarde presented at CRP/ in 2010. Archive, CRP/, Douchy-les-Mines.

43 Fred Klaba, Marie-Jo Noclain and Sylvie Dureuil, place des Nations, behind the CRP/.

45 Enlarger, darkroom number 1, Le Fresnoy, Tourcoing.

46 Automatic color processing machine Hoestert RA4, darkroom facilities, Le Fresnoy, Tourcoing.

47 Digital processor Lambda Durst 131, Le Fresnoy, Tourcoing.

48–49 Digital facilities, Le Fresnoy, Tourcoing.

Borinage

53 Grande Place, Quaregnon.

54 Charbonnage de Marcasse, Wasmes.

55 Hornu, rue de Binche.

56 Charbonnage de Marcasse, Wasmes.

57–59 Place du Levant, Levant de Mons, Bray.

61 Posters for European Parliament elections of May 25, 2014, Place du Levant, Levant de Mons, Bray.

63 Rue de Wasmes, site du Grand-Hornu, Hornu.

64–67 Site du Grand-Hornu, Hornu.

69 Sebastien Laurent, guide, MAC's, Musée des Arts Contemporains de la Féderation Wallonie-Bruxelles, site du Grand-Hornu, Hornu.

70 Exhibition *L'image suivante. Choix dans les collections de la Fédération Wallonie-Bruxelles*, MAC's, Musée des Arts Contemporains de la Féderation Wallonie-Bruxelles, site du Grand-Hornu, Hornu.

71 Patrizia Notarrigo, MAC's, Musée des Arts Contemporains de la Féderation Wallonie-Bruxelles, site du Grand-Hornu, Hornu.

73 Cafeteria, Site du Grand-Hornu, Hornu.

Monuments

75–92 Fosse d'Arenberg, Wallers.

93–102 Fosse 9–9bis, Oignies.

Mining Museum, Lewarde

105–124 Centre Historique Minier de Lewarde.

106 Mine reconstruction.

107 Anna Hembise, guide.

110 Diorama.

112 Museum shop.

115 Hospital diorama.

117 Daniel Francke, guide and former miner.

119 Museum shop.

120 Aurélie Bourgois, ticket sales.

121 Diorama.

Becoming heritage

129 Maison Syndicale du Minier, Lens.

130–131 Board room, Maison Syndicale du Minier, Lens.

132–135 Former printing facilities, Maison Syndicale du Minier, Lens.

136–138 Inventory, Maison Syndicale du Minier, Lens.

139 Serge Barruis, inventory, Maison Syndicale du Minier, Lens.

141–143 Screening of TV documentary series *Mémoires de la mine* (TF1, 1981), presented by director Jacques Renard. Organised by Mineurs du Monde, Faculté Jean Perrin, Université d'Artois, Lens.

145–147 Fosse d'Arenberg, Wallers.

149–150 Mission Bassin Minier, Fosse 9–9bis, Oignies.

151 Catherine O'Miel, Mission Bassin Minier, Fosse 9–9bis, Oignies.

152 Raphaël Alessandri, Mission Bassin Minier, Fosse 9–9bis, Oignies.

153 Marie Patou, Mission Bassin Minier, Fosse 9–9bis, Oignies.

154 Naima Maziz, Mission Bassin Minier, Fosse 9–9bis, Oignies.

155 Catherine O'Miel's office, dossier for the UNESCO world heritage declaration of former mining sites. Mission Bassin Minier, Fosse 9–9bis, Oignies.

156 Catherine Bertram, Mission Bassin Minier, Fosse 9–9bis, Oignies.

159–161 Archives du Monde du Travail, Roubaix.

163 Interior scene (c. 1867), historic theme galleries, Familistère Godin, Guise.

165 Renovation of central hall, Familistère Godin, Guise.

New/Old Economy

169–172 Fosse 9–9bis, Oignies.

175 Fosse d'Arenberg, Wallers.

176 Film set. Fosse d'Arenberg, Wallers.

177–178 Fosse d'Arenberg, Wallers.

181–188 La Plaine Images, Tourcoing.

191–195 Université de Valenciennes, campus at Aulnoy-les-Valenciennes.

197–203 Bombardier Transport SAS factory, Crespin.

205–213 Centre de Production Thermique, Bouchain.

Leisure

215 Flight HOP 3430 to Strasbourg, Lille airport.

216 Lille airport. Louvre-Lens advertisement in arrivals hall.

217 Noon, Euralille from the Pont de Flandres, Lille.

219-221 Loisinord, Noeux-les-Mines.

223 Fosse 9–9bis, Oignies.

224–225 Le Métaphone, Fosse 9–9bis, Oignies.

227 Memorial of 1974 mining catastrophe, Fosse 3 bis, Liévin.

228 Shopping mall, former site of Fosse n° 1, Liévin.

231–235 Louvre-Lens.

Renaissance?

Documentary Methods, Post-industrial Economies and Ideologies on Cultural Heritage

Panel discussion with Jacques Lemière, Jorge Ribalta and Hilde Van Gelder, presented and conducted by Muriel Enjalran.

CRP/ Centre régional de la photographie Hauts-de-France (formerly Nord-Pas-de-Calais), Douchy-les-Mines, December 5, 2015.

JL	Jacques Lemière
JR	Jorge Ribalta
HVG	Hilde Van Gelder
ME	Muriel Enjalran

ME: In the context of the exhibition *Renaissance* at the CRP/, I am thrilled to welcome Hilde Van Gelder and Jacques Lemière. They will be with us during this talk to exchange with Jorge Ribalta. They will refine our view of this project and shed light on the challenges raised in the exhibition, notably regarding the social, economic, political and ideological challenges of the "future of cultural heritage" for the industrial sites and landscapes of the region. Through the series of photographs in *Renaissance*, we will tackle the documentary question in the fields of visual culture and art and examine the ways in which it contributes to the debate on "cultural economies".

Before listening to you speak, Hilde and Jacques, about Jorge's work and the themes we briefly mentioned, could you, Jorge, look back on the series of photographs that surround us, the context in which they were made, the project's origins, how they fit in your work and concerns?

JR: I will talk about two things. First, what I try to do, in a kind of programmatic way; second, what you can see in the exhibition.

For the last ten years, my work has been an excavation of the unconscious implicit in the photographic apparatus and, more specifically, of

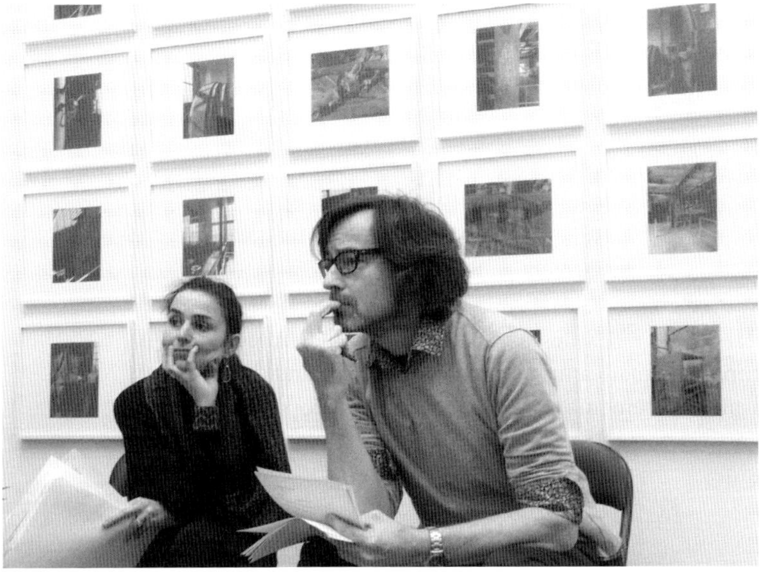

Muriel Enjalran and Jorge Ribalta during the panel discussion at the CRP/. Photo credit Quentin Derouet CRP/.

the *documentary idea*, an expression which I am borrowing from John Grierson in a non-literal way and which, to me, seems central to photography. My artistic approach consists in bringing the historical documentary idea to the present through a re-scenification of the three key moments in its origin and its evolution in the history of photography. These three moments are like geological layers in what I call the unconscious of the photographic apparatus.

The first moment took place around 1850. This is when the conditions for the emergence of a documentary episteme in photography emerged, even if the word or concept of the documentary as we've known it since the 1920s had not yet been coined. This was when photographic technology based on the positive/negative was established. It was, of course, the era of Blanquart-Evrard, who, to me, is the true father of photography as a public art. It was the birth of reproducibility, based on the combination of the collodion negative and the albumen print. This led to the first generation of photographic albums with the founding of Blanquart-Evrard's printing company in 1851. It was also the era of the first large-scale photographic surveys in history, particularly the *Mission Héliographique* in France, also in 1851. The historical simultaneity of the *Mission* and the beginnings of photographic printing was not a coincidence, it indicated the rise of a new public role of photography. The first conclusion after observing this deeper geological layer is that photography was born as an instrument of Nation State logics, for archival and propagandistic uses. Monuments and patrimony cannot be dissociated from state rationale, from the discourse of nationalism.

The second key moment took place around 1930, when the documentary discourse, as we know it now, emerged. While the function of the documentary predates this period, the birth of a discourse that specifically defined itself as documentary happened in 1926. One of the intellectual fathers of this idea, John Grierson, defined it as "a creative treatment of actuality". It is the inter-war period, the birth of mass media, the emergence of modern visual culture with photography and cinema taking a central role (magazines and newspapers illustrated with photography, compact format cameras such as the Leica and the Rolleiflex, the massification of cinema as public spectacle, propagandistic exhibitions, advertising and image-based technologies of opinion and persuasion, photography and cinema being incorporated in avant-garde art...). The documentary was a discourse that appeared precisely within this tension between art and the media. In this context, the documentary discourse was meant to represent the common man, the man of the streets or, more precisely and as William Stott said, the worker. My hypothesis is that the birth of the documentary discourse at that time had the mission of representing the new political protagonism of the working class in the era of the birth of mass democracy. One must understand that this was when the worker photography movement started, it was the era of Willy Münzenberg, of Soviet factography, of John Grierson's British documentary movement, of course, but it was also the era of the Farm Security Administration in the United States. During this period, photography was also theorised as allowing one to represent and understand social complexity. In that sense, the Russian Sergeï Tretiakov is a key figure for me. Consequently, the second observation that can be made when looking at this moment is that the emerging documentary form was an art form with a mission to represent work and workers.

The third moment is the 1970s. In a way, in my opinion, this was the birth of our time. It was a moment during which the documentary idea was re-signified as a tool for a critique of modernism (post-war institutional humanist modernism). This was when Allan Sekula and the San Diego group talked about the double task of "reinventing documentary" as a critique or a de-naturalisation of modernism. In the 1970s, there was a rereading of the 1930s, which for the first time went against post-war humanist immanence. With feminism, cultural studies, semiology, psychoanalysis and Marxism, photography definitively appeared as an ideological field, a space of conflict. I refer to the contributions of, say, Jo Spence in England and Martha Rosler in the United States, whose work produced a new micro-political understanding of representation.

In this context, a self-critical reflexivity had to be incorporated to photography to make implicit power relations visible, and to make one's own institutional inclusion relatively transparent and open to dispute. So the third assertion that can be made based on the observation of this period of the 1970s is that no one could hold the monopoly on photographic meaning any longer. And, in a certain way, more than a century after its birth, photography finally fulfilled its original promise of being a democratic art, an aesthetic reversal of bourgeois culture, as Molly Nesbit alleged in her study on Atget. The 1970s was a time when photography finally gained broad popular access. In this context, Jo Spence suggested that the aesthetic avant-garde could finally be appropriated in one's living room. This is an open discussion, but in the 1930s with worker's photography, this democratization had not taken place in the same way.

Therefore, based on this sort of historical excavation of these three constitutive moments of the documentary idea, I can formulate my artistic project: to pursue the modern documentary project, but in an self-reflexive way, of representing the field of culture, of cultural work, which is my work and the field in which I am inscribed.

Another general, programmatic question concerns the use of the exhibition space according to a certain notion of image saturation. There are close to 170 images in this exhibition. What does that mean? I offer two key ideas. The first is that this way of utilizing the exhibition space goes in the direction of producing what I call a "public photographic space". This has two meanings: an archaeological meaning and another polemical one. The archaeological meaning indicates a reference to the 1850 *salons*, the way in which photography appeared in the bourgeois public sphere in the artistic and industrial exhibitions of the time. It is a reference, or better, a response to Baudelaire's 1859 famous condemnation of photography, in the sense of advocating for the minor or popular uses of photography. It is a reference to the family album or to these various popular ways of using images on walls. I also think a lot, in Spain, of Flamenco culture clubs, who put up their own representations and their own genealogy on the walls. It is a very characteristic gesture when photography is used in a popular – not an artistic – way, and I want to update this in my work. And this brings me to the second meaning, the polemical one, which I mentioned: claiming these popular usages implies a critique of the tableau form, the history of painting model which has been dominant since the 1980s in occupying the institutional spaces of art with photography. Regarding photography, Jeff Wall may have best explained this idea, in a way that I respect: the bourgeois public space established itself through the collective contemplation of the large tableaux in the *Salons*. The tableau form materialised the bourgeois public sphere. Against the hegemony of the tableau form in the bourgeois artistic space, I defend photography as a "minor art form". By defending this "minor art", I want to be polemical in my way of highlighting the opposition of the bourgeois public space *versus* a plebeian space, a kind of working class counter-public sphere. There is therefore a struggle for public space. This is never a unified or peaceful concept, in the sense that it always involves some exclusions. I am therefore being polemical in my defence of photography's filiation with this sort of plebeian public sphere. And I use the word "polemical" because I'm aware that there is a democratic promise that is implicit in the bourgeois public sphere, and because I oppose this promise to that of a plebeian space highlighting the conflictual and dialectic nature of the public sphere. Call it class struggle if you prefer...

The second key idea, regarding the question of the public space being produced by photography, is the idea of the series, or of seriality being an epistemic condition for photography to be associated with the documentary idea of the 1930s. Tretiakov actually stated that seriality was precisely what allowed photography to represent and explain social complexity: the relation between the images is precisely what produces knowledge. I think that the notion of seriality, in the late 1920s, can be interpreted in at least three different ways. I suggest thinking about three different serial models from the late 1920s. The first is August Sander's model of "comparative photography", a term which Alfred Döblin coined in his introduction to Sander's book, *Antlitz der Zeit*, which says that Sander does a sort of "sociology in images". This was a serial model in the late 1920s. Another model is found in the iconographic relations in Aby Warburg's late work, *The Mnemosyne Atlas:* how can the relations between images, purely iconological relations, explain a certain history of artistic forms and of artistic culture? A third model of seriality at the end of the 1920s is Tretiakov's notion of an "extended photographic observation". Tretiakov says that in order for photography to represent this social complexity, the mobility of perception must be made visible, and he suggests a method of arranging images in groups and sequences which parallels photo-journalism methods and which can be associated with the photographic essay.

What I try to do is to update this documentary idea as being the result of an accumulation of historical sedimentations. I want to somewhat excavate the memory of the perception of photography and see how photography determines a way of thinking and a psychological mechanism.

These were the more general, or programmatic, questions. So, what can we see in the exhibition? *Renaissance* is the photographic observation of the reconversion process of a productive coal-based industrial economy into a cultural economy based on leisure and entertainment infrastructures and on the cultural industries of visual communication, media, digital culture and so on. Such a local process is not exclusive to the Nord-Pas-de-Calais region in France, where we are now. It is a global process, in which we, artists, art institutions, participate (even if we don't want to). A question mark could have indeed been added to the title of this project. And the Louvre-Lens quote at the entrance of the exhibition ["the Museum is at the forefront of a coal-mining basin in full renaissance"], which also gives the exhibition its title, should also bear a question mark. The question is: with industrial reconversion and what is known as "cognitive capitalism", are we really witnessing an economic, social and cultural "renaissance"? One of the characteristics of this process of economic reconversion is the use of propaganda and of cultural governance, of a rhetoric in which a certain social consensus is sought and in which interrogations on the economic model's negative social effects are hidden. *Renaissance* also happens to be, as some from the area might know, the title of the 2015 edition of the festival Lille 3000. The coincidence was completely unintentional. I learned about it later, after I had already named my project; and maybe Jacques will talk about all of that later.

So, the question is: will such a post-industrial economy, with all its grandiloquence, parallel the conditions that industrial post-war capitalism gave to the working classes? Eric Hobsbawn, whom I appreciate, called this post-war period the "golden age of capitalism", precisely because of its positive effects on the improvement of the working class. I therefore want to interrogate this reconversion process.

Even the sub-title, *Scènes,* which I used for the series, has a double meaning. Of course, scenes are views and images, but they are also moments in a theatrical representation. In that sense, the theatrical model is important in my work. I introduced a theatrical reference for the first time in a series I was working on while I was making this one, around the abdication of Charles V , king of Spain in the sixteenth century. I started using a term I like a lot: "documentary drama". In this context, Charles V is a useful figure to me. He serves as a sort of alter ego, a figure that allows me to do a tragi-comical duplication. Charles V is my post-modern Hamlet, a sort of pathetic Atlantes. I find that he constitutes a very useful tool to think about our current times.

The structure of the *Renaissance* series fits with this theatrical model and is organised in three acts. The first act is the most historical one: the pure observation of contemporaneity triggers a process of memory and fantasy. When we see the brand new Louvre-Lens, suddenly the rise of the European Nation State capitalist system in the sixteenth century comes to mind. The second act touches upon the question of "becoming a monument" or "becoming heritage": that strange moment when the factory or the mine closes and becomes a cultural object or a monument. And the third act is on the dialectic between work and leisure, one which is constitutive of the post-industrial economy and the promise of a world after work.

The *Renaissance* series starts where my other series on the abdication and death of Charles V, titled *Empire*, ends: namely with images of the portrait of Charles V's banker Anton Fugger on view at the Renaissance gallery in the permanent collection of the Louvre-Lens. These are two different images of the very same place and portrait. I come from Spain and my work is an attempt to identify and criticise the type of nationalism that has been internalised in the cultural sphere, and thus in the first sediment I mentioned before, which is part of the constitutive unconscious of documentary culture. And the Spain of Charles V is a very important reference to build this critique of Spanish nationalism since he is a highly mythicized figure representing the rise of the myth of the Spanish Empire. But he also represents the beginning of a modern European capitalist economy based on banking and public debt in the sixteenth century. That is what the Renaissance means historically, of course: it is the rise of the Modern era. It's therefore a key figure for those various reasons. After *Empire*, *Renaissance* is the second part of my trilogy in progress on Charles V, with the third part addressing America and the problem of colonisation.

Such a "documentary drama" is my way of reinventing the documentary idea today, as a method that is both historical and delirious. It aspires to produce a readable form for our times, one which is based on the photographic observation of the historical present's materiality and which must also account for the psychological aspects of the very act of observation.

Now Hilde is going to address the politics of documentary. Jacques will then talk about a potential critique of the cultural heritage imperative and the implicit ideological aspects of the evolution and potential future of the culture of industrial work.

HVG: In terms of the "critical condition" of current documentary photography, Jorge's project, which surrounds us here at the CRP/, is a remarkable example. The artist thought about the exhibition at length. It is essential for contemporary photography and these images merit our profound engagement. Each one asks to be unsealed in several layers. The spectator immediately experiences the invitation of taking his/her time with the installation, as it is designed specifically for this space. What initially sharply strikes us is the formal interplay between the black and white geometrical elements of the old floor, preserved in perfect condition, and the photos. This interaction between the images and the architecture of the space reinforces the visual intensity of the photographs on the walls. At the same time, we feel an intimacy when we find ourselves in this beautiful place. Jorge, you were talking a few minutes ago about a struggle for public space. It is interesting to think about this paradox in that a struggle for public space can happen by way of the delicate intimacy of an artistic project.

In regards to the fundamental question of the "documentary critique" in these very photographs, my impression is – and this could be a personal thought, Jorge – that you do not develop the usual documentary critique. It seems to me that you aim to go beyond Allan Sekula's artistic photographic project, a work which you often reference and which, in its "critical realism" already wasn't a documentary in the traditional sense of the term and tipped towards the visionary. With *Renaissance* you have succeeded in developing documentary photography that wants to be explicitly visionary. Of course your images have a "critical" approach towards their subjects, but the reflexive subtlety of your work definitely overcomes obvious reproach. You show and you even demonstrate elements of your "documentary critique" but you also look forward and you don't explicitly judge the past.

In this sense, it seems to me that your project, in its essence, aims to contribute towards building a reality of the present. You submerge analogue photography – quite an old medium today – in a renewal effort. To do this, you have developed a performative creation process asking the spectator to create, consider and think with you and in your wake. You suggest conceiving a common, shared reality, and you ask us to try imagining it collectively, inclusively. For the visitor, this approach leads to the sensation of there being openness, a "possible" way, shall we say. The spectator experiences strong sensations of what isn't there, what is not known yet. In its visionary aspect, your work announces and anticipates a future that is yet to come.

I would now like to articulate the broad impression of what happens inside your exhibition in more concrete terms.

The first area I want to address is the title: *Renaissance*. The great French historian of the Renaissance, Patricia Falguières, has magnificently linked concepts from the Renaissance to contemporary art. She suggests thinking of the Renaissance as a major historical event that also led to a major break at a certain point in history. According to her, the collective transformation of society caused by the Renaissance is something we are experiencing again today. We are going through a similar moment of profound upheaval. Some of the photographs present here in the space seem to connect with this idea. They even seem to confirm it visually, and therefore, they materialize it.

For example, in the case of climate change, the current situation being so dramatic, a certain "renaissance" would be needed at the level of global society in its entirety. It is a question tackled by your project – not directly, but by way of its intense focus on the region's cultural reconversion. You wonder to yourself: this cultural reconversion, will it be enough to affect durable social change on its own? When you came

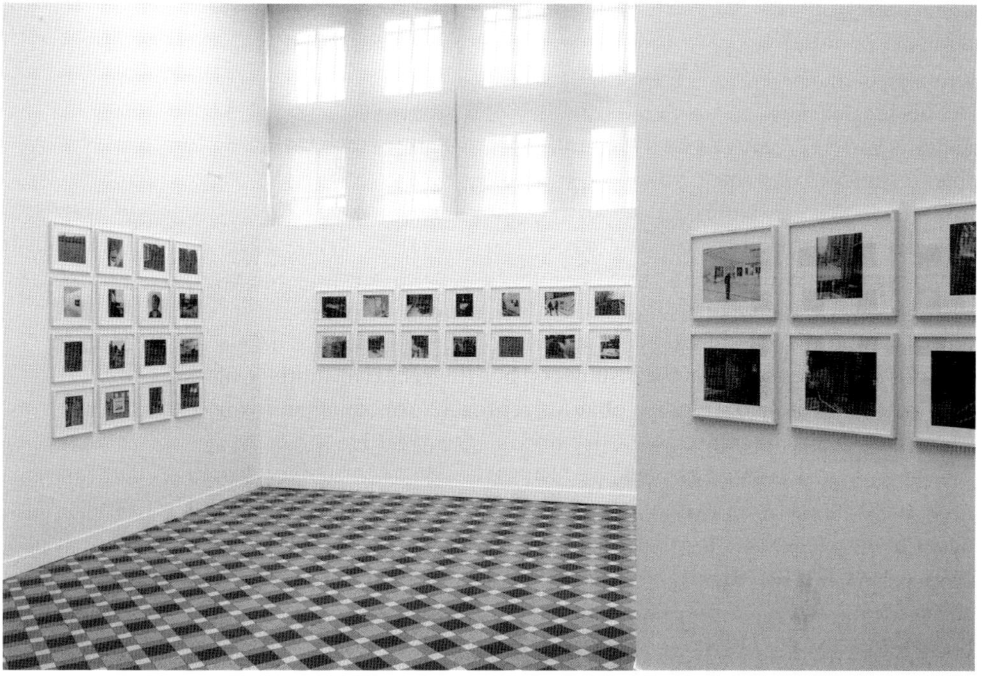

Exhibition *Renaissance* at the CRP/. Photo credit Quentin Derouet CRP/.

to Brussels to visit me to prepare for this discussion, Jorge, you were already raising this strong concern. You wondered what the residents of the region would think of your images. You were curious to see if you [the audience, the people of the region] recognised what Jorge as an artist wanted to communicate in his project. Some of the places he photographed may be familiar to you, or there might be other smaller elements that you recognise and that speak to you. It is precisely through these details that we can start seeing a larger history appear, and where the *Renaissance* project finds its current vitality and, even, its urgency.

I would like to end with a second line of thought that focuses on the four tools you, Jorge, have used in the series of photographs presented here. These tools are essential to your photographic strategy.

First, and we already said this: only analogue photographs are presented in *Renaissance*, so there are no digital images or interventions in this project. You have limited yourself to printing the images in a very traditional way, in a darkroom. On a theoretical level, the consequence is that we delve into theories of photography that are no longer very trendy, especially the theory of the "indexicality" of photography, mainly developed by Rosalind Krauss at the end of the last century. According to this theory, which has been very influential, photography is a physical and material trace of reality, the latter of which appears to be represented in the image. The process of developing the image becomes the mediator in this "indexical" mechanism. It is thanks to this physical process of development that the image can have a material relationship with the reality it shows and which is no longer there.

This idea of the trace, in which this connection between image and reality is fostered, leads us to the second tool, which is the relation to time. It seems that in the installation of the *Renaissance* series, co-temporalities – several simultaneous time frames – are presented. You, Jorge, have mentioned the eight "scenes" that you set up. These eight scenes translate into three temporalities which are different from each other but which nevertheless coexist. The first temporality is there, behind you: it is the historical time frame, the traces of the past. Then on the wall to your left and in front of you, there is the temporality of current times, the reconversion of mining sites transformed into cultural heritage sites. Finally, there is the temporality of the future, which is over there in the next room. The temporality of the past, as you say Jorge, is the "long duration" temporality of capitalism, a form of capitalism hailing from the Renaissance. Then there is the relatively slow temporality of the patrimony's reconversion. And finally, you make us face the hyper-rapid temporality of new sites, like the Louvre-Lens, being implanted – this can also be traumatizing. Amid coexisting temporalities, you invite us to think about the question of the "monument": why do we create monuments? Can they act as counterweights to the new temples of deregulated capitalism, all of which we see being erected everywhere around us?

The third tool, or element, is the portrait aspect, the physiognomy of people. What strikes me most in your image "scenes", especially in relation to what we just talked about on the question of monuments, is the fact that you included portraits. Due to the frontality of the portraits, it almost seems like people are suddenly talking to the viewer. This is striking, interesting, and adds to the narration of the sequences.

To finish, the fourth element, which I will touch on more briefly, consists in the connection between image and text. It is an interesting relationship because the text is present in the space but it is relatively hidden. What I mean is that it is not fixed to the wall but instead circulates as a separate brochure – which is to be read over the course of the visit, if you will. It was a conscious strategy on your part, Jorge. We could venture that this is linked to the fact that your work as an artist is fuelled by your work as a curator, as a theorist... Such wealth of experience was reflected in your deliberate decision to place the text slightly to the side, but nevertheless making it present enough for the audience. This creates a context for artistic engagement which aims to give information discreetly to the audience without sealing the relationship between the image and the text for it to "explain" everything. Your project, which is,

I'll say it again, visionary, consciously seeks to preserve a certain level of mystery. This is what I particularly appreciate.

JL: My contribution here will be of a different nature: I do not place myself in the realm of a critical outlook on photography, nor can I display previous work in the field of photography, even though I give photography a place in the seminar *"Images, sounds, social science"* which I started and lead at the University of Lille (at the *Espace Culture* of the *Cité scientifique* campus). In this seminar, after having invited other photographers, I invited Jorge in December 2014, for him to present his work, which he just now mentioned, on one of the periods of his "genealogy", the time of militant workers' photography in the inter-war period between the world wars.

From my position as a sociologist and as an anthropologist, and as I offer my point of view on the current moment of transformation in our societies, I might say things that could hurt people if they were to be misunderstood. I will say these things with tact and respect, and I want to point out from the outset that we are confronted with extremely contradictory and complex phenomena, to which we all participate. As Jorge said, from his position as an artist: "This cultural consumption, which is offered to us as the present in relation to a completed past and as a future, we as artists participate in it". In the same way, I include myself in what I am about to say on certain aspects of cultural policy. I do not claim to have a superior position, one from which some chosen people look and judge from above what is happening "below", where everything isn't wonderful. Indeed, while I am an observer of certain contemporary uses of culture, I am also involved in cultural programming (at *Citéphilo* in Lille, not necessarily only relating to film; at the Méliès *Nouveau Ciné-Club* in Villeneuve d'Ascq, which seems, to me, to be an initiative that is close in spirit to the offerings presented here at Douchy-les-Mines). I also participate in presentations on Portuguese cinema, which I study closely, and I am therefore an agent of initiatives instituted by public cultural policies (even though I try to be as independent as possible, as non-reliant on public money as possible and to keep a distance to remain free). Without wanting to be the police station's bad cop – the less sympathetic role, when others have nicer parts – it is nevertheless necessary to talk about certain things related to the usages of culture and heritage in our current society.

We exist in a determined moment of our society, and this moment is documented by the exhibition we are discussing. While the temporal sequence of this moment is documented, the photographer's investigation also leads us to places where a determined cultural and patrimonial policy has started to bear its mark on the region as a whole. This started in Lewarde in 1984, and the exhibition ends with an image of the Louvre-Lens, at the end of 2012. 30 years are represented, and we can now view this period with some distance. I would therefore like to start unfurling my thoughts with the idea that we are immersed in, involved in a new ideology: an "all cultural" ideology, one which crosses with the notion of "patrimonial machinery". I have borrowed this expression of "patrimonial machinery" from Henry-Pierre Jeudy, an author who wrote a very stimulating short book called *La machinerie patrimoniale.* (The editor may have been afraid of the word *machinerie* since, strangely, the book has two titles – this is a rare mistake in publishing: the front cover is titled *La machine patrimoniale,* which is a softer title than the more incisive *machinerie*, a term which is stronger in its critique. But, once the front cover turned, the opening title is *La machinerie patrimoniale*). As for the expression "*tout culturel*" ("all cultural"), it seems to fit in a line of thought supported by authors such as Alain Brossat, who published the book *Le grand dégoût culturel.* We should note that we are witnessing an invocation of "culture" as a totem, and, at the same time, this totemisation is linked to what one could call addictive patrimonization.

Culture invoked as a totem. We can take a look at the editorials published in municipal, departmental and regional communication bulletins (this is less true of the new Hauts-de-France regional presidency, which focuses its institutional discourse on access to employment and less on access to culture, unlike the previous Nord-Pas-de-Calais region presidency). The use of the word culture is never defined but is instead essentialized as a "totem" which refers back to the theme of rebuilding a social connection, one which we know was undone when the material industries disappeared. These industries employed workers alongside a whole host of employees and technicians and were equipped with all of the structures that go with the workings of factories.

We must mark the difference between art and culture. This difference must be highlighted for us not to lose sight of the stakes of the discussion. Art stems from gestures that transgress and that divide. Art is not straightforward and doesn't automatically unite audiences. Culture is art that has been accepted, consolidated, crystallized and that can be transmitted. In this sense, culture unites. You'll notice that culture is summoned to "rebuild social connections", to "re-vitalize an area", to "generate appeal…"; art is not summoned to do that. We talk about a "new cultural economy", of a "new cultural economy in the territories" just like we talk about the "knowledge-based economy" in universities, which may not be for the best. Just as "art" differs from "culture", the word "thought" differs from the world "knowledge". We can see here that it would be senseless to talk about a "thought economy", just as we wouldn't thing of promoting a "new economy of art". This is therefore a "cultural economy" and it is what's changing in the fields of leisure and culture, just as it is changing in universities: all mayors of mid-sized cities in France have wanted their own universities. In the same way, most mayors want their museum or multimedia library to be set in a former factory.

Addictive Patrimonization. Of course, respect is due to the workers' pride in the industrial past as well as in the skills and ways of doing things that are linked to that past. Patrimonization transforms mining and industrial work places into places of leisure, performance centres, culture spaces, museums, art schools, universities... just as Jorge has documented in his photographs.

We had already seen this dynamic at work in the field of agriculture and rural life. The great transformations of agriculture led to the disappearance of small peasantry and farms. As a young sociology student, I experienced this in the mainland of Brittany in the 1970s, alongside poor farmers who were confronted with the bulldozer that was authoritarian farmland re-parcelling: these were farmers who were still living, who hadn't modernised and were not mechanised much (therefore weren't in debt!). They were trying to survive on their land and to resist unequal re-parcelling, which they knew would benefit the farmers who had joined the ranks of modern agriculture, as it was called. And yet, not far from them, the museum of past agriculture already existed: as if the dead were seizing the living.

This can be a terrifying prospect in the process of patrimonization: the dead seizing the living. Life is still present, but certain laws and transformational plans threaten it... the farmers I mentioned weren't against all transformation, nor were they against all re-parcelling principles. They said: we agree with amiable re-parcelling, we will discuss exchanging our land around a table, for a fair exchange. The land regrouping they wanted wasn't the authoritarian one desired by the State, Agricultural Chambers and productivist farming unions. These small farmers were alive, but an agriculture that was known as being modern, for which the bulldozers of re-parcelling are a good metaphor, were advancing to swallow them. The museum of traditional peasantry already existed. The small farmers were alive, but their ways of life and work were already moving from the farm to the museum.

I also found this "death seizing the living" aspect in my investigations in the territories of northern France and southern Belgium, places that are documented in Jorge's work. This is more visible with the patrimonization of places that are not places of industrial work but rather living spaces connected to the former industrial workplace. When work disappeared, when the factory went bankrupt or was liquidated – since very profitable factories are liquidated for reasons of globalised economic calculations – when the factory closes, when work disappears, when people can no longer go to the factory in the morning for employment, people nevertheless live there, and they see patrimonization establish itself with all of its multiple uses, as patrimonization also has a stake in the housing of workers and miners.

One of the photographs exposed by Jorge is of particular interest to me, it is one of the pavilions of the *Familistère* of Guise, where the *Association du Capital et du Travail*, founded by Jean-Baptiste Godin, an industrialist, was born amid utopian socialism. This attempt died out in 1968 (a ruse of history!), after many trials and tribulations during the 19th and 20th centuries ... The factory (manufacturing Godin stoves) got separated from the housing complex, which had never been Godin's plan, on the contrary. And people still lived in the apartments of the *Palais social* pavilions. And then the residents were chased out to make way for a new type of museum project, one of European scope (a large museum for the tourism and cultural economy). This transformed the place, making it into a celebration of utopia, one which was as monumental as it was pedagogical – and global. There had previously been a small museum consisting in a small reconstituted worker's apartment on the ground floor. The museography was judged as obsolete, but this small museum nevertheless had the merit of respecting popular housing in a site where comfort and functionality had been a golden rule for Godin, its founder. Before the residents were chased out of their homes to make way for a large museum which, unlike the old one, was to take up the entire space, protest signs and flyers were found inside the *Familistère* and featured slogans against the museum seizing the apartments. Of course, Godin's associative and collective project had been dead for a long time by then, but the last *familistère* members, or the descendants of *familistère* members were strong, they lived there, they tried to resist, they were displaced from a site that was now completely being turned into a museum.

Another example of such ideological, social and political stakes – not linked to the heritage conservation of buildings this time – lies in the destruction and reconstruction of lived-in spaces. There, memory is summoned to legitimize destruction, the latter of which remains traumatizing to the people who live there. This is an urban example and relates to foreign workers' homes built in the 1960s. These collective homes are threatened by a law that aims to transform these spaces into "social housing", thus preventing residents from having a (partially) community-based lifestyle, (and by "community", I don't mean "ethnic community ghettos" at all) in the sense that this lifestyle is based on collective uses and cooperation (for example, the use of shared kitchens). One of the major ways in which this law has been applied has been through the physical destruction of these group homes under the pretext that the buildings were getting old (they had willingly been left un-renovated for a long time); in this case, the destruction isn't only physical but it is also legal, since the living spaces and living rules are bound to considerably change in the newly built spaces. This touches upon the command people have over their ways of living, and the relationships between them. And since such an operation of destruction and reconstruction must be legitimized, "in-residence" photographers (a ruse of the lexicon this time, instead of a ruse of history!) are called to collect the memory of what is about to be destroyed. These urban renovation operations, as they are called, are linked to the State's distrust of group homes and of the self-defined community life that develops within them. During these operations, the artist in residence collects the memory of the life that had developed in these spaces (these artistic ventures are included in the operational budget of the ANRU, *Agence Nationale de la Rénovation Urbaine*). The photographs of the residents are then grouped and displayed on the construction site's tarpaulins. Sometimes, (there was a case of this in Lille), cynicism was prevented at the last minute when a few people stood up and said: "*This can't be, you're not going to display the photographs of these African and Arab workers on the construction site's tarpaulins, workers who lived in this home for 30 years, in a building you're destroying and rebuilding while expelling some of these people in the process*". Some of these ANRU operations took place in social housing projects where the photographs of residents who spent their lives in the tower blocks were displayed on the construction site's gates. And a variety of other examples could be cited.

In such instances, the position of the artist resembles that of the journalists who were embedded – and, therefore, relatively safe on the battlefield – in the American tanks of Bush's army during the invasion of Iraq, in a deal with the office that controlled military information. We are led to examine the responsibility of artists in such situations. This is why I like the fact that Jorge visibly placed the expression "*intervention de l'art*" ("art intervention") under the exhibition's overall title. In other words, will art intervention be conscious and critical of these stakes, or will it be complicit? This is all the more important as the new patrimonial ideology is always at risk of producing clichés, since the violence of work and the violence of social relations are eliminated in the sites that are turned into patrimony. Entering these places is indeed made much easier once the true memory of such violence has been eliminated.

I am very impressed by Jorge's work in this way, as it is so artistically and politically refined, in the noblest sense of the term. The work is conducted without dogmatism – because, of course, new activities must begin in these areas, people must find a renewed sense of hope of being able to stay where they are.

To finish, I'll share echoes of my investigations on the popular feelings around patrimony.

The first echo takes place in Auchel, in the western part of the former coal-mining area of Nord-Pas-de-Calais. The other echo takes place in

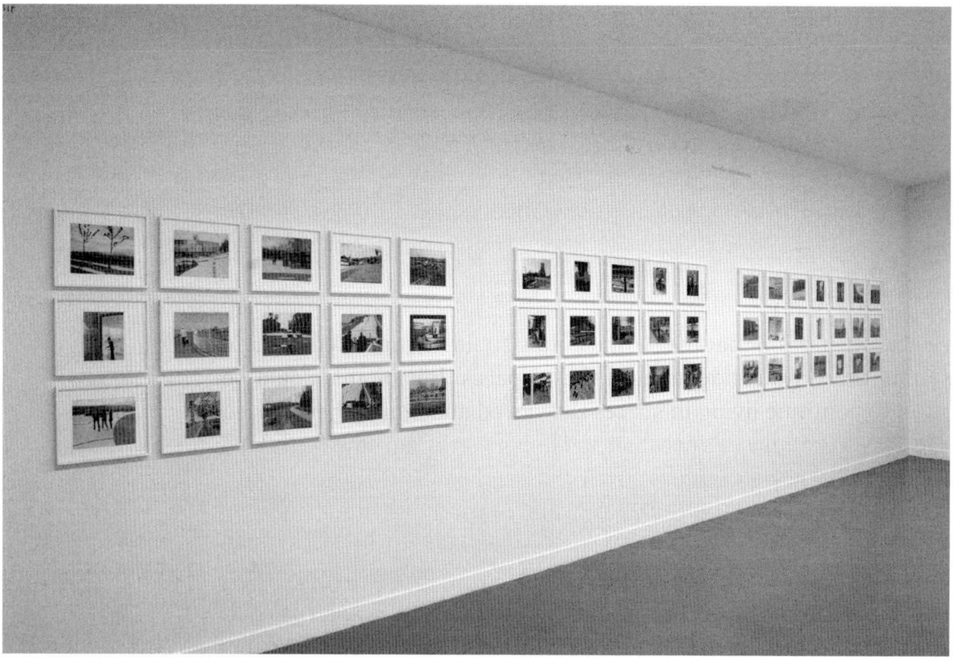

Exhibition *Renaissance*. Photo credit Quentin Derouet CRP/.

the central part of the region, in Lens and Liévin (the Louvre-Lens museum was only a project at the time), a former coal-mining region in which I was conducting an ethnological study which wasn't centred on the issue of popular housing but rather on a particular practice wherein people from this northern region displayed objects which could be seen from the street on the window sills of their homes' ground floors. According to shared social codes, the passer-by's gaze should not penetrate the intimate space of the house through a window. But due to the presence of these objects (sometimes a collection of objects) exhibited as if they were in a store window, one was allowed to stop a little bit, to look, at the risk of the gaze entering the domestic intimate sphere. The paradoxically porous nature of the domestic and public spaces led us to become interested in this popular practice of placing objects on window sills. Several things came out of this study, starting with the ease with which we were able to converse with the residents despite the fact that we did not hide our act of photographing the displayed window sill objects: the gesture of taking photographs was not an artistic stance but a technical one, with the view of compiling an inventory and coming into contact. When we entered the houses, not only did we sometimes find personal museums for which the window sill objects were only a part of the whole, but our conversations in the workers' row houses – the *corons* – often led the residents to want to testify about their living conditions and against the decision made by the landlord (in this case, the SOGINORPA company) of destroying the row houses. Some people were left living alone between boarded-up houses in these threatened *coron* neighbourhoods. A popular view of patrimony, which deflected the official discourse, could then be heard. "*Look how nice our little houses are, they would need to be renovated a bit. And actually, since so many young couples are looking for houses, why should these be destroyed? And it's our heritage*", said a woman who then pointed to the small rebuilt houses that were to replace the *coron* homes at the other end of a large patch of lawn: "*Have you seen those Mickey Mouse houses?*" It was "our (*coron*) heritage" versus "the Mickey Mouse houses", and there was the notion that the heritage sites could be protected without upsetting the ways residents used these spaces. Inhabitants could be allowed to continue living in spaces they loved and which they carried with them – through the quotidian and living use of the houses – a long working-class and popular memory.

These are, as we said, complex and contradictory phenomena, especially when it comes to the resilience of these deindustrialised territories, something to which we cannot be indifferent. From the perspective of the residents of the former coal-mining area, the Louvre-Lens project could seem bizarre: during our study in Liévin – at the exact place where Jorge photographed the preserved headframe next to the commercial centre, a woman who lives opposite the headframe told us: "*In any case, the Louvre-Lens isn't for us. And, besides, it's not even for our children. They're not going to take in our children to work there*". A few years later, in December, on the evening of *Sainte-Barbe* day, at 7:00 pm, when the Louvre-Lens museum was open to the people for the first time, the same people who hadn't been invited to the official inauguration in the morning, incredible lines had formed (some visitors didn't enter until 9:00 pm!) of people who all had very different histories and experiences of the museum, and who were all coming to see the beauty of the space: "*It's beautiful! It doesn't feel like we are in Lens*", said one woman to her friend, at one of the control areas at the entrance of the Louvre-Lens. People came to see Beauty land in the town of Lens like a UFO, but they also came to appropriate this beauty for themselves. It was baffling to those of us who had heard so much scepticism in the years following the announcement of the project and during the museum's construction. Because, although people were aloof in their judgment of the cultural policies that corresponded to the Louvre-Lens project, these people were still present. The Louvre-Lens director had intelligently decided that the permanent exhibition would be free of charge and accessible to all, while the temporary exhibition would cost the price of a ticket. In his novel *L'Assommoir*, the scene described by Zola of the wedding party's visit to the Louvre comes to mind, and it was important to see what this museum invitation would lead to, in long form, after the Sunday meal: on Sunday, at 3:00 pm, in the museum's *Galerie du Temps*, the attendance of working-class families was incredible. Popular uses of the space could therefore take place wherein the saying *"the Louvre-Lens isn't for us"* was not confirmed. This, of course, doesn't take away the need for finer and more specific surveys on the attendance and uses of the museum for both the temporary and permanent exhibitions... But I nevertheless wanted to offer these observations to our discussion.

ME: What you said about the Louvre-Lens is very interesting. It is probably too early to make an assessment, but we are starting to hear criticism being formulated regarding the economic promises that were made, especially for the local area, of employment being provided and of a virtuous circle forming when Lens residents would work at the Louvre-Lens, etc. We hear this as a source of criticism: after the initial feelings of openness and excitement to wake the town up and make it move towards something new, something may not have fully happened in this move from one economy to another. Perhaps because it was too ambitious, even at the scale of this museum, to think that we could contribute to the question of employment.

JL: I can perhaps add to this a few observations from the other side of the Franco-Belgian border, in the Walloon region, where Jorge also conducted photographic work. In my case, these observations stem from a study trip I took with master's students at former coal-mining sites such as Le Grand Hornu near Mons and Le Bois du Luc, near La Louvière. At Bois du Luc, paternalist industrialists added a model company town to the mine since its creation in the 19th century. The model city was designed based on the paradigm of employer control. It consisted in bosses preventing their workers from getting corrupted by emerging socialist ideas in the nearby town, La Louvière. Since the mine closed at the end of the 20th century, an interesting museum has opened and is composed of a very dynamic and committed team. But a paradox was troubling its curator, a woman who is the daughter of a coal miner from the Maghreb region: while the company town benefited from the "UNESCO World Heritage Site" label, its residents were being evicted. *"What should I do to help the renters of the mining complex keep their housing? How can we call out the city hall and the social landlord and ask them how they can do such a thing in a company town that has the UNESCO label?"*, she asked with a certain amount of anxiety.

The other example is Le Grand Hornu. Le Grand Hornu was a coal mine since the end of the 18th century. After 1815, it was overseen by a former officer in the Napoleonic wars. He implemented an architectural system that was reminiscent of the earlier system of the Royal Saltworks at Arc-et-Senans. It was inspired by the same idea of combining the workplace with the living space, the latter of which included leisure and education areas such as a theatre, and followed the same principle of employer rationality. What is striking in the patrimonization that took place at Le Grand Hornu – a site destined to become a ruin if a passionate architect hadn't taken an interest in it – is that the site was saved by giving it the function of a contemporary art space, thus instigating a complete break with was had been a place of work in the 19th century (and a place for the mine's administration). The workers' housing complex surrounding this space isn't included in the guided tours of the site, thus suspending the transmission of the place's history. This is yet another version of this ongoing process of patrimonization: the partial patrimonization of a site, which dismembers it in a way, while also giving it a new contemporary cultural function that is totally separate from the site's history. Jorge's photographic exhibition documents this brilliantly.

Renaissance ?

Méthodes documentaires, économies post-industrielles et idéologies du patrimoine culturel

Débat avec Jacques Lemière, Jorge Ribalta et Hilde Van Gelder, présenté et animé par Muriel Enjalran.

CRP/ Centre régional de la photographie Hauts-de-France (formerly Nord-Pas-de-Calais), Douchy-les-Mines, 5 décembre, 2015.

JL	Jacques Lemière
JR	Jorge Ribalta
HVG	Hilde Van Gelder
ME	Muriel Enjalran

ME : J'ai le plaisir d'accueillir autour de l'exposition *Renaissance* au CRP/, Hilde Van Gelder et Jacques Lemière qui vont nous accompagner le temps de cette rencontre dans un échange avec Jorge Ribalta pour enrichir notre approche de ce projet et nous éclairer sur les enjeux qu'elle aborde ; notamment sur les enjeux sociaux, économiques, politiques et idéologiques du « devenir patrimoine » des sites et paysages industriels de la région. À travers la série de photographies *Renaissance,* nous aborderons ainsi la question documentaire dans le champ de l'image et de l'art et sa contribution au débat sur les « économies de la culture ».

Avant de vous entendre, Hilde et Jacques, à propos du travail de Jorge et des thèmes que nous venons d'évoquer rapidement, pourriez-vous Jorge revenir sur la série de photographies que nous avons autour de nous, leur contexte de réalisation, la généalogie de ce projet, et l'inscription et la place de cette série dans votre œuvre, ses enjeux ?

JR : Je vais parler de deux choses. La première, de ce que j'essaye de faire, d'une façon un peu programmatique. Deuxième chose : de qu'est-ce que vous pouvez voir dans l'exposition.

Mon travail ces dix dernières années, consiste en une excavation de l'inconscient du dispositif photographique et particulièrement de *l'idée documentaire,* une expression que j'emprunte de façon non littérale à John Grierson et qui me semble être une idée centrale pour la photographie. Mon opération artistique consiste à porter au présent cette idée documentaire historique au travers d'une re-scénification des trois moments clefs dans son origine et évolution dans l'histoire de la photographie. Ces trois moments sont des sortes de strates géologiques de ce que j'appelle l'inconscient du dispositif photographique.

Le premier c'est autour de 1850. C'est le moment dans lequel se donnent les conditions pour l'émergence d'une épistémè documentaire dans la photographie, même si le mot ou concept documentaire tel que nous le pensons à partir des années 1920 n'est pas encore produit. C'est le moment d'implantation de la technologie photographique fondée sur le positif/négatif. C'est évidemment l'époque de Blanquart-Evrard, qui pour moi est le vrai père de la photographie en tant qu' art public. C'est l'époque de la naissance de la reproductibilité, fondée sur la combinaison du négatif au collodion et le tirage à l'albumine, qui produit la première génération d'albums photographiques à partir de la fondation de l'imprimerie de Blanquart-Evrard en 1851. C'est aussi l'époque des premières grandes campagnes photographiques de l'histoire, particulièrement la Mission Héliographique en France, aussi en 1851. N'est pas par hasard la simultanéité historique de la Mission et le début de l'imprimerie, indicatifs du nouveau rôle public de la photographie. Le premier constat de l'observation de cette plus profonde strate géologique, c'est que la photographie naît comme instrument pour l'archive et la propagande au service de l'État Nation. Les monuments, le patrimoine, sont indissociables de la logique étatique, du discours du nationalisme.

Le deuxième moment clef c'est autour de 1930, quand le discours documentaire spécifiquement dit naît. Même si la fonction du documentaire est antérieure, la naissance d'un discours qui se définit spécifiquement comme documentaire est en 1926. Un des pères intellectuels de cette idée, John Grierson, le définit comme « le traitement créatif de l'actualité ». C'est l'époque de la période d'entre-guerres, de la naissance des médias de masse, de l'éclosion de la culture visuelle moderne fondée sur la centralité de la photographie et du cinéma (les revues et la presse illustrées photographiquement, les appareils de petit format comme les Leica et Rolleiflex, la massification du cinéma comme spectacle public, les expositions propagandistes, la publicité et les technologies d'opinion et de persuasion par l'image, l'incorporation de la photographie et du cinéma à l'art d'avant-garde...). Le documentaire est un discours qui apparaît justement dans cette tension entre l'art et les médias. La fonction, dans ce contexte, du discours documentaire est de représenter l'homme commun, l'homme de la rue, ou plus exactement, et comme le dit William Stott, le travailleur. Mon hypothèse est que, justement, la naissance du discours documentaire à cette époque a la mission de représenter le nouveau protagonisme politique de la classe ouvrière dans cette époque de la naissance de la démocratie de masse. Il faut comprendre que c'est aussi l'époque du mouvement de la photographie ouvrière, de Willy Münzenberg, de la factographie soviétique, du mouvement documentaire britannique de John Grierson, évidemment, mais c'est aussi l'époque de la Farm Security Administration aux États-Unis. C'est aussi un temps qui théorise que la photographie permet de comprendre la complexité sociale, et dans ce sens, le russe Sergeï Tretiakov est une figure clef pour moi. En conséquence, le deuxième constat qui vient en observant ce moment est que le documentaire est un art qui émerge avec la mission de représenter le travail et les travailleurs.

Le troisième moment, ce sont les années 1970. D'une certaine façon, à mon avis, c'est la naissance de notre époque. C'est un moment de ré-signification de cette idée documentaire comme un instrument de critique de la modernité (disons, la modernité humaniste institutionnelle de l'après-guerre). C'est l'époque où Allan Sekula et le groupe de San Diego parlent de la double opération de « réinventer le documentaire » en tant que critique ou dénaturalisation de la modernité. C'est dans les années 1970 qu'il y a une relecture des années 1930, qui pour première fois est contraire à l'immanence humaniste de l'après-guerre. Avec le féminisme, les études culturelles, la sémiologie, la psychanalyse et le marxisme, la photographie apparaît définitivement comme un champ idéologique, comme un espace de conflit, avec les contributions de Jo Spence en Angleterre ou Martha Rosler aux États-Unis. C'est-à-dire qu'il y a tout un discours pour une nouvelle lecture micro-politique de la représentation et, dans ce contexte, la photographie doit incorporer une dimension autocritique et rendre visibles les rapports de pouvoir qui sont implicites, rendre visible sa propre inscription institutionnelle. Donc, le troisième constat à partir de l'observation de cette période des années 1970 est que personne ne peut plus avoir le monopole de la signification de la photographie et, d'une certaine façon, plus d'un siècle après sa naissance, finalement la photographie accomplit sa promesse originelle d'un art démocratique, un revers esthétique de la culture bourgeoise, comme le signalait Molly Nesbit dans son étude d'Atget. C'est dans ces années 1970 que finalement la photographie a un accès populaire généralisé et c'est dans ce contexte que Jo Spence que l'on s'approprie l'avant-garde esthétique dans la salle à manger. On peut en discuter, mais dans les années 1930, avec la photographie ouvrière, cette démocratisation ne s'est pas produite de la même façon.

Alors, de cette sorte d'excavation historique de ces trois moments constitutifs de l'idée documentaire, se formule, pour moi un certain projet artistique : la continuation du projet documentaire moderne, mais d'une façon autoréflexive, pour représenter le champ culturel, le travail culturel, qui est mon travail et qui est le champ dans lequel je m'inscris.

Une autre question générale, programmatique, c'est la question de l'usage de l'espace d'exposition avec une certaine idée de saturation d'images. Dans l'exposition il y a presque 170 images. Qu'est-ce que cela veut dire ? Je vous propose deux idées clefs. La première c'est que cette façon d'utiliser l'espace d'exposition va dans la direction de produire ce que j'appelle « un espace photographique public ». Cela a deux sens : un sens archéologique et un autre polémique. Le sens archéologique, cela veut dire qu'il y a ici une référence aux salons de 1850, la manière dont la photographie est apparue dans l'espace public bourgeois dans les expositions artistiques et industrielles de l'époque. C'est une référence, ou mieux, une réponse à la fameuse condamnation de la photographie faite par Baudelaire en 1859, dans le sens de faire une revendication de l'usage mineur ou populaire de la photographie. C'est une référence à l'album de famille ou à ces différentes façons d'usage populaire des images sur les murs. Je pense surtout, en Espagne, aux associations de culture « flamenco », qui mettent sur le mur toutes leurs propres représentations, toute leur propre généalogie. C'est un geste très caractéristique de cet usage populaire, pas artistique, de la photographie que je veux réactualiser dans mon travail. Et ça me porte au deuxième sens, polémique, dont je parlais : revendiquer ces usages populaires implique une critique au modèle du tableau, le modèle d'histoire de la peinture, le modèle dominant depuis les années 1980 pour occuper l'espace institutionnel de l'art avec la photographie. En ce qui concerne la photographie, c'est Jeff Wall qui a peut-être mieux expliqué cette idée, d'une façon que je respecte : l'espace public bourgeois se constitue sur la contemplation collective des grands tableaux dans les Salons. La forme tableau matérialise la sphère publique bourgeoise. Devant cette forme tableau dans l'espace artistique bourgeois historique, je propose une défense de la photographie comme un « art mineur » et de rendre visible au travers de la défense de cet « art mineur », d'une façon polémique, cette opposition entre espace public bourgeois *versus* espace public plébéien. Il y a donc une lutte pour l'espace public, qui n'est jamais unifié ou universel. Je défends donc de façon polémique la filiation de la photographie avec cette sorte de sphère publique plébéienne. Et je dis polémique parce que je suis conscient de la promesse démocratique de la sphère publique bourgeoise quand j'oppose à cette promesse celle d'un espace plébéien que veut souligner la constitution conflictuelle et dialectique de la sphère publique. Enfin, lutte de classes, si vous voulez le poser comme ça…

La deuxième idée clef par rapport à cette question de l'espace public produit par la photographie, c'est l'idée de série ou de sérialité comme une condition épistémique de la photographie associée à cette idée documentaire qui vient des années 1930. Justement, c'est Tretiakov qui affirme que c'est justement la sérialité qui permet à la photographie d'expliquer et de rendre visible la complexité sociale : c'est justement le rapport entre images qui produit le savoir. Cette idée de sérialité, à la fin des années 1920, selon moi, peut s'interpréter au moins de trois façons différentes. Je vous propose de penser à trois modèles sériels différents de la fin des années 1920. Le premier c'est le modèle d'August Sander, le modèle de la « photographie comparative », un mot introduit par Alfred Döblin dans l'introduction du livre *Antlitz der Zeit*, qui dit que Sander fait une sorte de « sociologie en images ». Döblin introduit le concept de « photographie comparée ». Voici un modèle de sérialité de la fin des années 1920. Un autre modèle, ce sont les relations iconographiques d'Aby Warburg avec le projet de la fin de sa vie, l'*Atlas Mnémosyne :* comment les relations entre images, les pures relations iconologiques, expliquent une certaine histoire des formes artistiques et de la culture artistique ? Un troisième modèle de sérialité de la fin des années 1920, c'est le modèle de Tretiakov, l'idée d'une « observation photographique élargie ». Tretiakov dit que pour que la photographie puisse représenter cette complexité sociale, il faut rendre visible la mobilité de la perception, et il propose une méthode d'images en groupes et séquences pas loin des méthodes du photojournalisme, que l'on peut aussi associer à l'essai photographique.

Bref, ce que j'essaye de faire, c'est de réactualiser cette idée documentaire comme constituée par une accumulation de sédimentations historiques, et de faire cette sorte d'excavation de la mémoire de la perception de la photographie et de voir comment la photographie détermine une forme de pensée et un mécanisme psychique.

Telles étaient les questions un peu plus générales, ou programmatiques. Alors, qu'est-ce que l'on peut voir dans l'exposition ? *Renaissance* est l'observation photographique du processus de reconversion d'une économie productive industrielle, fondée sur l'exploitation du charbon, à une économie de la culture fondée sur la consommation culturelle et sur les industries culturelles de la communication, l'audiovisuel, la culture numérique, les loisirs... Un processus qui n'est pas seulement local de la région Nord-Pas-de-Calais en France où nous sommes maintenant, il s'agit d'un processus général, dans lequel nous, les artistes, les institutions de l'art, participons (que nous le voulions ou non). Le titre de ce projet pourrait effectivement avoir un point d'interrogation. Et la citation du Louvre-Lens dans l'entrée [*Le musée est la figure de proue d'un bassin minier en pleine renaissance*], qui donne le titre à l'exposition, devrait aussi avoir un point d'interrogation. La question est : avec la reconversion industrielle et le dit « capitalisme cognitif », sommes-nous vraiment devant une « renaissance » économique, sociale et culturelle ? Une caractéristique de ce processus de reconversion économique est l'usage d'une propagande ou d'une gouvernance culturelle, d'une rhétorique qui cherche un certain consensus social, qui ne rend pas visible les questions sur les effets sociaux négatifs du modèle. *Renaissance,* c'est aussi, comme le savent peut-être ceux qui viennent de la région, le titre de Lille 3000 pour son édition de 2015, ce que j'ai appris plus tard et par hasard, après avoir mis mon titre au projet, et peut-être que Jacques va parler de tout ça après.

Alors, la question c'est : est-ce que cette économie post-industrielle, avec toute sa grandiloquence, va proportionner les conditions pour les classes populaires qu'avait produit le capitalisme industriel de l'après-guerre (cette période que l'historien Eric Hobsbawn, que j'aime beaucoup, a appelé « l'âge d'or du capitalisme ») ? L'époque qui a vu le plus grand progrès des classes populaires dans l'histoire de l'humanité. Je veux donc interroger ce processus.

Le sous-titre *Scènes* que j'ai utilisé pour la série a encore un double sens. Évidemment, *Scènes* sont les vues, les images, mais sont aussi les moments d'une représentation théâtrale. Dans ce sens, le modèle théâtral dans mon travail est important. J'ai introduit une référence théâtrale pour la première fois dans une série que j'ai faite auparavant, en même temps que je faisais celle-ci, autour de l'abdication de Charles Quint, Roi d'Espagne au XVIème siècle, dans laquelle j'ai commencé à utiliser un mot que j'aime beaucoup : « drame documentaire ». Dans ce contexte, Charles Quint est une figure qui me sert comme une sorte d'alter ego, une figure qui me permet un dédoublement tragicomique. Charles Quint c'est mon Hamlet post-moderne, une sorte d'Atlante pathétique, que je trouve très utile pour penser notre époque.

La structure de la série *Renaissanc*e, en accord avec ce modèle théâtral, s'est organisée en trois actes. Le premier est le plus historique, qui vient de ce mécanisme de l'observation de la contemporanéité qui met en fonctionnement tout un dispositif de mémoire et de fantaisie. Quand on regarde le tout nouveau Louvre-Lens on pense aux débuts du système capitaliste dans les états-nation en Europe au XVIème siècle. Le deuxième acte porte sur cette question de « devenir monument » ou « devenir patrimoine » : cet étrange passage quand l'usine ou la mine arrête, et devient culture ou monument. Et le troisième acte est sur la dialectique entre travail et loisir, aussi constitutive de l'économie post-industrielle et la promesse d'un monde après le travail.

Cette série commence au point où finit ma série sur l'abdication et la mort de Charles Quint qui a pour titre *Empire.* Cette série finissait avec une image du portrait du banquier Anton Fugger au Louvre-Lens, une image de la même situation et du même lieu mais différente de l'image qui marque le début de *Renaissance.* Je viens d'Espagne. Mon travail est une tentative d'identifier et de critiquer le nationalisme intériorisé en tant qu'inconscient constitutif de la culture documentaire et du champ culturel, et l'Espagne de Charles Quint est une référence très importante pour construire cette critique du nationalisme espagnol. Charles Quint est une figure hyper-mythifiée, le début du mythe de l'Empire Espagnol. Mais c'est aussi le début d'une économie capitaliste européenne moderne fondée sur la banque et la dette publique au XVIème siècle. La Renaissance dans son sens historique, bien sûr. Alors, c'est une figure clef pour plusieurs raisons. *Renaissance* est la deuxième partie de ma trilogie en cours sur Charles Quint, dont la troisième aborde l'Amérique et le problème de la colonisation.

Ce « drame documentaire », c'est ma manière de réinventer le documentaire aujourd'hui comme une méthode à la fois historique et délirante qui aspire à produire une forme de lisibilité de notre temps à partir de l'observation photographique de la matérialité du présent historique, et qui doit aussi rendre compte des aspects psychiques de cette observation même.

Maintenant Hilde va nous parler de la politique du documentaire et puis Jacques nous parlera d'une possible critique de la patrimonialisation et des aspects idéologiques implicites dans ce devenir de la culture du travail industriel.

HVG : Par rapport à la « condition critique » de la photographie documentaire actuelle, le projet de Jorge qui nous entoure ici au CRP/ est un exemple remarquable. Très longuement réfléchie par l'artiste, son exposition, fondamentale pour la photographie contemporaine, mérite que nous nous engagions profondément avec ces images. Chacune d'elles demande d'être descellée en plusieurs couches. Conçue spécifiquement pour cet espace, le spectateur ressent immédiatement l'invitation de prendre son temps avec l'installation. Dans un premier moment ce qui frappe très fort est le jeu formel entre les éléments géométriques noirs et blancs du sol ancien, préservé en parfait état, et les photos. Cette interaction entre images et architecture de l'espace renforce l'intensité visuelle des photographies qui y sont attachées aux murs. En même temps on ressent une intimité quand on se retrouve dans ce bel endroit. Jorge, tu parlais tout à l'heure d'une lutte pour l'espace public. C'est intéressant de penser à ce paradoxe qu'une lutte pour l'espace public peut passer par l'intimité délicate d'un projet artistique.

En ce qui concerne donc la question fondamentale de la « critique documentaire » de ces photos-mêmes, mon impression est – et ça peut être une pensée personnelle, Jorge – que tu ne développes pas une critique documentaire classique. Il me semble que tu vises à passer même au-delà du projet artistique photographique d'Allan Sekula, œuvre à laquelle tu fais beaucoup référence et qui dans son « réalisme critique » déjà n'était pas documentaire dans le sens classique du terme non plus mais balançait également vers le visionnaire. Avec *Renaissance* tu réussis à développer une photographie documentaire qui se veut être en même temps explicitement visionnaire. Bien sûr tes images ont une approche « critique » par rapport au sujet qu'elles traitent, mais la subtilité réflexive de ton travail surmonte largement le flagrant reproche. Tu montres, voire démontres des éléments de ta « critique documentaire » mais tout autant tu regardes vers l'avant et tu ne juges pas explicitement le passé.

Dans ce sens, il me semble que ton projet dans son essence même veut contribuer à la construction d'une réalité du présent. Tu submerges la photographie analogique, pourtant un médium fort vieux aujourd'hui, dans un effort de renouvellement. Pour cela tu développes un processus de création qui est performatif et qui invite le spectateur à créer, penser et réfléchir ensemble avec toi et dans ta lignée. Tu nous proposes de concevoir une réalité commune, partagée, et tu nous demandes d'essayer de l'imaginer collectivement, solidairement. Au niveau du visiteur, cette approche suscite des sensations d'une ouverture, d'un « possible » on pourrait dire. Le spectateur ressent des fortes sensations de ce qui n'est pas là, de ce qui n'est pas encore connu. Dans son aspect visionnaire, ton travail annonce et anticipe ainsi sur un futur encore à venir.

Je voudrais maintenant articuler cette impression plus générale de ce qui se passe à l'intérieur de ton exposition dans des termes un peu plus concrets.

Le premier axe que je veux aborder est le titre : *Renaissance*. La grande historienne française de la Renaissance, Patricia Falguières, a magnifiquement su lier des concepts qui datent de la Renaissance à l'art contemporain. Elle propose de penser la Renaissance comme un évènement historique majeur qui a aussi causé, à un certain moment de l'Histoire, une forte césure. La transformation collective de la société qui a été causée par la Renaissance, nous sommes, selon sa conviction à elle, en train de la revivre aujourd'hui. Nous traversons, pour ainsi dire, un moment similaire de bouleversement profond. Certaines des photos présentes ici dans l'espace semblent connecter avec cette idée, voire même la confirmer visuellement et donc la concrétiser.

Par exemple au niveau du changement climatique mondial, la situation actuelle est si dramatique qu'il faudra une certaine « renaissance » au niveau de la société mondiale toute entière. C'est une question que ton projet soulève non pas directement mais tout autant en se focalisant de manière intense sur la reconversion culturelle de la région. Tu te demandes : est-ce que cette reconversion culturelle à elle seule sera vraiment suffisante pour réaliser un changement social durable ? Quand tu t'es rendu à Bruxelles pour me rendre visite en guise de préparation de cette conversation, Jorge, tu soulevais déjà cette forte inquiétude. Tu te posais la question sur ce que les habitants de la région même allaient penser de tes images. Tu étais curieux de voir si vous [le public, les gens de la région] alliez reconnaître ce que Jorge en tant qu'artiste désirait communiquer dans son projet. Il y a sans doute des lieux qu'il a photographiés et qui sont connus pour vous, ou d'autres petits éléments que vous reconnaissez et qui vous parlent. C'est justement à travers ces détails que l'on peut commencer à voir une histoire plus grande et où le projet *Renaissance* obtient toute sa vitalité actuelle, voire son urgence.

Je voudrais terminer avec un deuxième axe qui focalise sur quatre outils que tu utilises, Jorge, dans la série de photos présentées ici. Ces outils sont essentiels à ta stratégie photographique.

Premièrement et on l'a déjà dit : dans *Renaissance* il n'y a que de la photographie analogique, donc pas d'image ni d'intervention numériques dans ce projet. Tu te restreins à un processus de tirage des images de façon très classique, en chambre noire. Au niveau théorique, la conséquence est que l'on entre un peu dans des théories de la photographie qui ne sont plus très à la mode aujourd'hui, surtout la théorie de « l'indexicalité » de la photographie, principalement développée par Rosalind Krauss à la fin du siècle dernier. Selon cette théorie, qui a joui d'une très grande influence, la photographie est une trace physique et matérielle de la réalité qui se révèle comme représentée dans l'image. Le médiateur de ce mécanisme « indexical » est le processus de développement de l'image. C'est grâce au procédé physique de son développement que la photo entretient une relation matérielle avec la réalité qu'elle montre et qui maintenant n'est plus là.

Cette idée de trace qui nourrit le lien entre image et réalité nous mène au deuxième outil, qui est celui du rapport au temps. Il semble que dans l'installation de la série *Renaissance* il y a une présentation de co-temporalités, de plusieurs temps simultanés. Toi-même, Jorge, tu parles de huit « scènes » que tu avais mises en place. Ces huit scènes se traduisent en trois différentes temporalités qui néanmoins coexistent. La première temporalité, elle est là, derrière vous : c'est la temporalité historique, ce sont les traces du passé. Puis au mur à votre gauche et devant vous, il y a la temporalité d'aujourd'hui, de la reconversion des sites des mines qui se sont transformés en sites du patrimoine culturel. Enfin il y la temporalité du futur, qui est plutôt là, dans l'autre pièce à côté. La temporalité du passé, tu dis Jorge, c'est la temporalité de la « longue durée » du capitalisme, un capitalisme qui nous est parvenu depuis la Renaissance. Ensuite il y a la temporalité assez lente de la reconversion du patrimoine. Finalement tu nous confrontes à la temporalité hyper-rapide et parfois aussi traumatisante de l'implantation de nouveaux sites, comme le Louvre-Lens. Dans cette coexistence de temporalités, tu nous invites à réfléchir sur la question du « monument » : pour quoi créons-nous des monuments ? Sont-ils capables de fournir un contrepoids aux nouveaux temples du capitalisme dérégulé que nous voyons être érigés un peu partout autour de nous ?

Le troisième outil ou élément, c'est l'aspect portrait de la physionomie des gens. Ce qui me frappe très fort dans tes « scènes » d'images, surtout en relation avec ce qu'on vient d'aborder sur la question des monuments, c'est que tu y intègres des portraits. Tout à coup, le spectateur voit des gens qui lui adressent presque la parole tellement la frontalité du portrait est frappante, intéressante, et s'ajoute à la narration des séquences.

Pour terminer, le quatrième élément, sur lequel je vais également être plus brève, concerne la relation entre image et texte. Elle est très intéressante parce que le texte est là, dans l'espace, mais il est relativement caché. C'est-à-dire qu'il n'est pas attaché au mur mais circule seulement comme brochure séparée – à lire en cours de parcours de visite si vous voulez. C'est une stratégie consciente de ta part, Jorge. On pourrait soulever l'hypothèse que ceci est lié au fait que ton travail d'artiste est nourri par ton travail de commissaire d'exposition, de théoricien… Cette richesse on la voit reflétée dans ta décision délibérée d'avoir le texte un peu à côté mais quand-même suffisamment présent pour le public. Cela crée un contexte d'engagement artistique qui vise à fournir discrètement au public de l'information sans pour autant vouloir bétonner la relation entre l'image et un texte qui viendrait à tout « expliquer ». Ton projet, visionnaire, je le répète une dernière fois, cherche consciemment à préserver une certaine énigme. C'est cela que j'apprécie particulièrement.

JL : Mon propos ici sera autrement situé : je ne me situe pas dans l'espace du regard du critique de la photographie, ni ne peux revendiquer une antériorité de travail sur la photographie, même si je donne une place à la photographie dans le séminaire « *Images, sons, sciences sociales* » que j'ai fondé et que j'anime à l'Université de Lille (à l'Espace Culture du campus de Cité scientifique), séminaire où, après d'autres photographes, j'ai invité Jorge, en décembre 2014, pour y présenter son travail, qu'il vient d'évoquer, sur une des périodes de sa « généalogie », la période de la photographie militante ouvrière de l'entre-deux-guerres mondiales.

De ma position de sociologue et d'anthropologue, portant son regard sur ce moment actuel de la transformation de nos sociétés, je vais peut-être dire des choses qui pourraient fâcher si elles étaient mal comprises. Je vais les dire avec tact et respect, en soulignant d'emblée, que nous sommes placés devant des phénomènes extrêmement contradictoires, et complexes, phénomènes auxquels nous participons tous, comme, de sa position d'artiste, Jorge l'a dit : « *Cette consommation culturelle qui nous est proposée comme présent par rapport à un passé révolu, et comme devenir, nous, artistes, nous en participons* ». De même, ce que je vais dire sur certains traits des politiques culturelles, moi-même, je m'y inclus. Je ne revendique donc aucune position de surplomb, dans laquelle quelques heureux élus pourraient regarder et juger d'en haut ce qui serait un « en bas » où tout ne serait pas formidable. Car, observateur de certains usages contemporains de la culture, je suis engagé moi-même dans des programmations culturelles (à *Citéphilo* à Lille, cinématographiques ou non ; au Nouveau Ciné-Club du Méliès à Villeneuve d'Ascq, initiative proche dans son esprit, je pense, de propositions qui ont existé ici à Douchy-les-Mines) comme dans des présentations de cinéma portugais, que j'étudie de près, je suis donc un acteur d'initiatives prises dans les politiques publiques de la culture (même si je m'y efforce d'être le plus indépendant possible, le moins dépendant de l'argent public, et d'y garder une distance pour rester libre). Sans être le méchant du commissariat – celui qui a le rôle pas sympathique, quand d'autres ont un rôle plus avenant –, il faut quand-même dire les choses sur des usages de la culture et du patrimoine qui sont faits dans notre société présente.

Nous sommes dans un moment déterminé de notre société, et l'exposition que nous commentons documente ce moment. Elle en documente la séquence temporelle, en même temps que l'enquête du photographe nous mène dans des lieux où une politique culturelle et patrimoniale déterminée a commencé à marquer son empreinte dans toute une région : cela commence à Lewarde, en 1984, et l'exposition se termine sur une image du Louvre-Lens, donc à la fin de 2012. Ce sont déjà 30 années, sur lesquelles on peut maintenant porter un regard avec recul. Je voudrais alors déplier mon propos à partir de cette idée que nous baignons, que nous sommes impliqués dans une nouvelle idéologie : une idéologie du « tout culturel », qui se croise avec une « machinerie patrimoniale ». Cette expression de « machinerie patrimoniale », je l'emprunte à un auteur qui s'appelle Henry-Pierre Jeudy qui a écrit un petit livre tout à fait roboratif qui s'appelle *La machinerie patrimoniale* (l'éditeur a peut-être eu peur du mot *machinerie*, car bizarrement l'ouvrage a deux titres, erreur rare dans l'édition : la page de couverture dit *La machine patrimoniale*, titre un peu plus soft que « machinerie », plus incisif, plus puissant dans la critique, mais, la couverture passée, l'ouverture du livre porte le titre *La machinerie patrimoniale*). L'expression de « tout culturel » peut, elle, s'inscrire dans une ligne de réflexion dans laquelle on trouverait des auteurs comme Alain Brossat, qui a publié l'ouvrage *Le grand dégoût culturel*. Retenons donc qu'on assiste à une invocation de la « culture » en totem, et qu'en même temps cette totémisation s'articule à ce qu'on peut appeler une patrimonialisation addictive.

La culture invoquée comme totem. Regardons les éditoriaux des bulletins de communication municipale, départementale ou régionale (c'est moins vrai avec la nouvelle présidence de la région Hauts-de-France, qui met au centre de son discours institutionnel la question de l'accès à l'emploi, et moins celle de la culture comme le faisait la précédente présidence de la région Nord-Pas-de-Calais). Cet usage du mot culture, sans jamais qu'il soit défini, usage qui l'essentialise au contraire comme « totem », le réfère à la thématique de la reconstitution du lien social, lien social dont on sait qu'il a été défait quand est disparue l'industrie matérielle, qui employait des ouvriers et, autour des ouvriers, tout un ensemble d'employés et de techniciens, avec toutes les structures qui entourent le fonctionnement des usines.

Notons la différence entre art et culture, qui est importante à relever si on veut ne pas perdre les enjeux de la discussion. L'art procède de gestes qui transgressent, et qui divisent. L'art ne va pas de soi et ne réunit pas d'emblée les publics. La culture, c'est de l'art qui a été accepté, consolidé, qui a été cristallisé, et qui peut être transmis. En ce sens, la culture réunit. Notons que c'est la culture qui est convoquée pour *« reconstituer le lien social »*, pour *« redynamiser un territoire »*, pour y *« créer de l'attractivité … »*, et non l'art. On parle d'une « nouvelle économie culturelle », de « nouvelle économie culturelle dans les territoires » comme, à l'université, peut-être pas pour le meilleur, on parle d' « économie de la connaissance ». Comme « art » se distingue de « culture », le mot « pensée » se distinguera du mot « connaissance » et on voit bien qu'il serait insensé de parler d'« une économie de la pensée », de même qu'on ne parlera pas de promouvoir « une nouvelle économie de l'art ». Il s'agit bien d'une « économie culturelle ». C'est ce qui change dans le champ du loisir et de la culture, comme cela change dans le champ universitaire : tous les maires des villes moyennes de France ont voulu leurs universités, comme à peu près tous les maires veulent leur musée dans l'ancienne usine et leur médiathèque.

La patrimonialisation addictive. Bien sûr, il y a le respect dû à ce qui fait la fierté ouvrière d'un passé industriel, des savoir-faire qui y étaient liés, des manières d'être qui y étaient attachées. La patrimonialisation transforme des lieux de travail minier ou industriel, comme Jorge le documente photographiquement, en lieux de loisir, lieux de spectacles, lieux de culture, musées, écoles d'art, universités ….

On a déjà vu cela à l'œuvre concernant le travail agricole et la vie rurale, dans les grandes transformations de l'agriculture qui ont fait disparaître la petite paysannerie et les fermes de l'agriculture paysanne. Jeune étudiant en sociologie, je l'ai éprouvé, dans les années 1970, aux côtés de paysans pauvres qui devaient faire face aux bulldozers du remembrement autoritaire des terres bocagères, au centre de la Bretagne : ils étaient encore des paysans vivants qui, bien que non modernisés, peu mécanisés (et donc non endettés !) essayaient de survivre sur leurs terres et de résister à ce remembrement inégalitaire dont ils savaient qu'il allait bénéficier à des paysans qui, eux, avaient pris le train de l'agriculture dite moderne. Or, non loin d'eux, il y avait déjà le musée de l'agriculture du passé : comme si le mort saisissait le vif.

C'est un aspect qui peut être terrifiant, dans les processus de patrimonialisation : le mort saisit le vif. Il y a encore du vivant, mais il y a des lois et des logiques de transformation qui le menacent ... Les paysans dont je parle n'étaient pas contre toute transformation, ni contre tout principe de remembrement. Ils disaient : on veut bien un remembrement amiable, on va échanger nos terres en discutant, autour d'une table, d'un juste échange. Le remembrement qu'ils voulaient, ce n'était pas celui, autoritaire, que voulaient l'État, les Chambres d'Agriculture et les syndicats agricoles productivistes. Ils étaient vivants ; une agriculture dite moderne, dont les bulldozers du remembrement étaient une bonne métaphore, avançait pour les avaler ; et il y avait, déjà-là, le musée de la paysannerie ancienne. Ils étaient vivants, mais il fallait que leur mode de vie et de travail passe de la ferme au musée.

J'ai retrouvé cette « saisie du vif par le mort » lors d'enquêtes dans ces territoires du nord de la France et du sud de la Belgique que le travail photographique de Jorge documente. C'est plus visible quand on observe la patrimonialisation de lieux qui ne sont pas des lieux de travail industriel mais des lieux d'habitat articulés à l'ancien lieu de travail industriel. Quand le travail a disparu, que l'usine a fait faillite ou qu'elle a été liquidée – puisqu'on liquide des usines très rentables pour des raisons de calculs économiques mondialisés – si l'usine est fermée, si le travail a disparu, si on ne peut plus aller le matin à l'usine pour embaucher, des gens néanmoins habitent là, qui voient la patrimonialisation s'installer avec tous ses poly-usages, car l'habitat ouvrier ou minier d'hier représente des enjeux de patrimonialisation.

Une photographie m'intéresse beaucoup parmi celles que Jorge expose, celle d'un des pavillons du Familistère de Guise, où l'Association du capital et du travail fondée par l'industriel Jean-Baptiste Godin, née dans les idées du socialisme utopique, s'est éteinte en 1968 (ruse de l'histoire !), après des vicissitudes enjambant 19ème et 20ème siècles ... On a alors séparé l'usine (celle des poêles Godin) de l'habitat, ce qui n'était pas le projet de Godin, bien au contraire. Et les appartements des pavillons du Palais social sont restés habités. Puis les habitants en ont été chassés pour qu'advienne un projet de musée, de type nouveau et de gabarit européen (un grand musée de la nouvelle économie touristique et culturelle), transformant les lieux en célébration monumentale autant que pédagogique – et mondiale – de l'utopie. Il y avait jusqu'alors un petit musée, réduit à la reconstitution d'un petit appartement ouvrier du rez-de-chaussée, avec une muséographie qui a été jugée obsolète, mais ce petit musée avait comme pourtant grand mérite d'avoir respecté le logement populaire dans un site où le confort et la fonctionnalité de cet habitat était pour son inspirateur, Godin, une règle d'or. Avant que les habitants ne soient chassés de leurs logements pour que ce musée ancien laisse place au grand projet sur l'ensemble du site, on trouvait des pancartes et des petites affichettes de protestation, à l'intérieur du Familistère, contre la reprise des appartements par le musée. Certes, le projet associatif et collectif de Godin était mort, depuis longtemps déjà, mais eux, les derniers « familistériens », ou descendants des « familistériens », étaient vifs, ils habitaient là, ils ont essayé de résister, ils ont été délogés d'un site désormais complètement muséifié.

Prenons un autre exemple de ces enjeux idéologiques, sociaux, politiques qui ne concernent pas que les situations de conservation patrimonialisée du bâti mais aussi celles de destruction et reconstruction de lieux habités, avec, alors, la convocation d'une mémoire de légitimation d'une destruction toujours traumatisante pour les gens qui les habitent. Cet exemple est urbain, et concerne les foyers de travailleurs étrangers construits dans les années 1960 et désormais visés par une loi qui empêche les résidents de ces foyers, par leur transformation en « résidences sociales », de connaître une vie (partiellement) communautaire (« communautaire », pas du tout « communautariste ») en ce qu'elle est fondée sur des usages du collectif et de la coopération (par exemple, l'usage de cuisines collectives). Un des moyens majeurs d'application de cette loi est de détruire physiquement ces foyers, sous prétexte de leur vieillissement (mais ils furent longtemps et volontairement laissés sans rénovation) ; la destruction n'est pas alors que physique, mais aussi juridique, par modification considérable des espaces de vie et des règles de vie à l'intérieur de ce qui sera reconstruit. Est alors atteinte la maîtrise que les gens ont sur leur façon d'habiter, et ce qu'elle dit des rapports entre eux. Et comme il faut légitimer l'opération de destruction et reconstruction, on fait souvent appel à un photographe "en résidence" (ruse, cette fois, non de l'histoire, mais du lexique !) pour qu'il recueille la mémoire de ce qu'on va détruire. La dite opération de rénovation urbaine, liée à une méfiance de l'État vis-à-vis de ces lieux et de la vie communautaire auto-décidée qui s'y développe, est accompagnée par le recueil, par l'artiste en résidence (il y a un budget dans les opérations de l'ANRU, Agence Nationale de la Rénovation Urbaine, pour ces opérations artistiques) de la mémoire de la vie qui s'y développait. D'où le recueil des photographies des habitants, qu'on va voir sur les bâches du chantier de la destruction. Quelquefois (on a en a connu un cas à Lille), ce cynisme a été empêché en dernière minute quand quelques-uns se sont levés pour dire : « *Ce n'est pas possible, vous n'allez pas mettre sur les bâches du chantier la photographie de ces ouvriers africains ou maghrébins qui habitent dans le foyer depuis 30 ans, foyer que vous allez détruire et que vous allez reconstruire en en expulsant une partie d'entre eux* ». On a connu ces opérations de l'ANRU dans des cités où les photos d'habitants qui ont passé toute leur vie dans des barres vouées à la destruction se sont retrouvées sur des panneaux du chantier, et on pourrait en multiplier les exemples.

La position de l'artiste ressemble alors à celle des journalistes *embedded*, embarqués – et ainsi relativement protégés, sur le champ de bataille – dans les chars américains de l'armée de Bush dans la guerre d'invasion de l'Irak, dans un deal avec le bureau du contrôle de l'information militaire. Cela questionne la responsabilité des artistes dans de telles situations. C'est la raison pour laquelle j'aime que, sous le titre général de son exposition, Jorge, dispose, en bonne place, l'expression « intervention de l'art ». Autrement dit : l'intervention de l'art va-t-elle être consciente, et critique, de tels enjeux, ou en sera-t-elle complice ? C'est d'autant plus important que la nouvelle idéologie patrimoniale est toujours au risque de la production de clichés car elle élimine, des sites qu'elle patrimonialise, la violence du travail et la violence dans les rapports sociaux. C'est en effet quand on a éliminé la mémoire vraie de cette violence qu'il est beaucoup plus facile d'aller dans ces lieux.

Le travail de Jorge, de ce point de vue, est d'une finesse artistique et politique, dans le sens le plus noble du terme, qui m'impressionne. C'est un travail mené sans dogmatisme – car, bien évidemment il faut que les territoires redémarrent avec des nouvelles activités, il faut que les gens retrouvent de l'espoir de pouvoir rester dans leur pays.

Des échos d'enquête, pour terminer, sur le sentiment populaire face à cette question du patrimoine.

Le premier écho se situe à Auchel, dans la partie occidentale de l'ex-Bassin Minier du Nord-Pas-de-Calais, l'autre écho dans sa partie centrale, à Lens et Liévin, ex-Bassin minier où je menais alors– et le Louvre-Lens n'était encore qu'un projet – , une enquête ethnologique qui n'était pas centrée sur les enjeux du logement populaire, mais sur la pratique des gens du Nord de placer des objets sur leur appui de fenêtre du rez-de-chaussée, objets qui sont visibles de la rue. Le regard du passant ne doit pas, selon le code social partagé, entrer, par la fenêtre de la maison de rue, dans l'intimité de la maison. Mais la présence de ces objets exposés comme dans une vitrine (et quelquefois de collections d'objets), autorise à s'arrêter un peu, à regarder, au risque que le regard entre dans l'intimité domestique. C'est cette paradoxale porosité entre espace domestique et espace public qui nous conduisait à nous intéresser à cette pratique populaire de placer des objets à la fenêtre. L'enquête disait plusieurs choses, à commencer par la facilité d'entrer en dialogue avec les habitants, sachant que nous ne cachions pas notre action de photographier les objets disposés aux fenêtres : geste de photographier non d'un point de vue artistique, mais technique, dans une logique d'inventaire, et geste pensé pour entrer précisément en contact. Entrant dans les maisons, non seulement on y trouvait quelquefois un musée personnel dont les objets à la fenêtre n'étaient qu'une pointe avancée, mais le dialogue, dans les corons, menait souvent à leur volonté de témoigner sur leurs conditions d'habitat, et contre les décisions du bailleur (en l'occurrence, la société SOGINORPA) de détruire des corons, des gens restant seuls à vivre entre des maisons murées, dans les corons ainsi menacés. Et on pouvait alors entendre une conception populaire du patrimoine qui venait en conjurer le discours officiel : « *Regardez, nos petites maisons, elles sont bien, il faudrait les rénover un peu. Et puis, il y a tant de jeunes couples qui cherchent des maisons, pourquoi les détruire ? Et puis c'est notre patrimoine » ; disait une femme qui désignait les petits pavillons de la reconstruction qui allaient remplacer les corons, à l'autre bout grand terrain gazonné : « Vous avez vu ces maisons de Mickey ? ». C'était « notre patrimoine (des corons) » versus « les maisons de Mickey »*, avec cette idée qu'on pourrait protéger un patrimoine bâti sans bouleverser les usages que les habitants en ont, et leur permettre de continuer à vivre dans des espaces qu'ils aiment et qui sont porteurs, dans un usage quotidien et vivant, d'une mémoire ouvrière et populaire ancienne.

Ce sont, on l'a dit, des phénomènes complexes et contradictoires, notamment sous le critère de la résilience de ces territoires désindustrialisés, critère auquel on ne peut être indifférent. Le projet du Louvre-Lens, du point de vue des habitants de l'ex – Bassin minier, pouvait apparaître comme un projet bizarre : dans cette enquête à Liévin – exactement là où Jorge a photographié le chevalement qui a été conservé, à côté du centre commercial. Une dame qui habite juste en face de ce chevalement nous dit : « *De toute façon, le Louvre-Lens, c'est pas pour nous. Et puis, c'est même pas pour nos enfants. C'est pas nos enfants qu'ils vont prendre pour travailler là-dedans* ». Quelques années plus tard, en décembre, le soir de la Sainte-Barbe, à 19 heures, quand le musée du Louvre-Lens fut pour la première fois ouvert au peuple, qui n'était pas invité à l'inauguration officielle du matin, ce furent des files d'attente incroyables (beaucoup n'entrèrent qu'à 21 heures) de gens, très différents dans leur histoire comme dans leur expérience propre du musée, qui venaient voir la beauté des lieux : « *C'est beau ! On ne se croirait pas à Lens* », disait une dame à son amie, à un contrôle à l'entrée du Louvre-Lens. Les gens venaient voir la Beauté débarquée comme un ovni à Lens et en même temps s'appropriaient cette beauté. C'était bluffant, pour qui avait entendu tant de scepticisme dans les années de l'annonce de projet puis du chantier du musée, car, même si l'opération Louvre-Lens correspondait à toutes ces politiques culturelles jugées avec tant de distance par les gens du peuple, ces gens étaient là. Et en cet hiver-là – le directeur du Louvre-Lens ayant imaginé, dans sa grande intelligence, que si l'exposition temporaire était payante, l'exposition permanente serait gratuite et libre d'accès –, et d'autant qu'on se souvient de la scène que décrit Emile Zola dans son roman *L'Assommoir* de la visite de la noce au musée du Louvre, il valait la peine d'aller voir ce que donnerait cette invitation au musée, au long cours, après le déjeuner dominical : le dimanche, à 15h, la fréquentation des familles populaires était formidable, dans la Galerie du Temps du musée. Il pouvait donc y avoir des usages populaires du lieu qui ne validaient pas la rengaine : « *Le Louvre-Lens, c'est pas pour nous* ». Ceci, bien sûr, ne dispense pas d'enquêtes fines et spécifiquement centrées sur la fréquentation et sur les usages du musée, dans l'exposition temporaire comme dans l'exposition permanente ... Mais je souhaitais proposer ces observations à notre discussion.

ME : C'est intéressant ce que vous disiez par rapport au Louvre-Lens. C'est peut-être tôt pour faire un bilan, mais c'est vrai qu'on entend déjà des critiques se formuler à l'encontre des promesses économiques, notamment pour un bassin local qui devait être porteur d'emploi et créer un cercle vertueux où les habitants de Lens allaient pouvoir travailler aussi au sein du Louvre-Lens, etc. Ça c'est l'une des critiques que l'on entend : après cet appel d'air et cet engouement pour réveiller cette cité et la faire basculer vers autre chose, finalement il y a peut-être aussi sur ce passage de l'économie quelque chose qui n'a pas pu être réalisé parce que c'était peut-être trop ambitieux même à l'échelle d'un musée, de pouvoir penser qu'on pourrait contribuer à cette question de l'emploi.

JL : Je pourrais peut-être ajouter quelques observations faites cette fois de l'autre côté de la frontière franco-belge, en Wallonie, où s'est aussi dirigé le travail photographique de Jorge, observations issues pour moi d'un voyage d'études, avec des étudiants de Master, sur des sites anciennement miniers comme Le Grand Hornu, près de Mons, et Le Bois du Luc, près de La Louvière. Au *Bois du Luc*, les patrons paternalistes adjoignirent à la mine, quand elle fut créée au 19ème siècle, une cité modèle complètement conçue dans le paradigme du contrôle patronal. Il s'agissait bien pour les patrons de retenir leurs ouvriers d'aller se faire corrompre dans la ville voisine, La Louvière, par les idées socialistes montantes. On y trouve aujourd'hui, après la fermeture de la mine, à la fin du 20ème siècle, un intéressant musée, avec une équipe très dynamique et très engagée, mais un paradoxe troublait sa conservatrice, fille de mineur venu du Maghreb : la cité ouvrière bénéficiait d'une labellisation « Patrimoine mondial de l'UNESCO » et en même temps on en expulsait des habitants. « *Comment pensez-vous que je doive faire si je veux aider les locataires de la cité minière à garder leur logement ? Comment interpeller la mairie ou le bailleur social et leur demander comment ils peuvent faire cela dans une cité minière labellisée UNESCO' ?* », me demandait-elle avec une certaine angoisse.

L'autre exemple, c'est Le Grand Hornu. Le Grand Hornu fut une mine depuis la fin du 18ème siècle, dirigée après 1815 par un ancien officier des guerres napoléoniennes qui y établit un dispositif architectural pouvant évoquer celui, plus précoce, de la Saline royale d'Arc et Senans, et inspiré par la même idée de la fusion entre le lieu de travail et le lieu d'habitat, ce dernier incluant des lieux de loisir et d'éducation, comme le théâtre, et sans conteste animé par le même principe de rationalité patronale. Ce qui frappe dans la patrimonialisation opérée au Grand Hornu – site qui a bien failli disparaître dans la ruine, si un architecte ne s'était pas passionné pour lui – , c'est qu'on l'a sauvé en lui donnant la fonction d'un lieu d'art contemporain, de sorte qu'on a institué une coupure complète entre ce qui était au 19ème siècle le lieu du travail (et de l'administration de la mine) et le dispositif des habitats ouvriers qui l'entourent, qui n'est pas inclus dans les visites guidées du site, laissant donc la transmission de son histoire en jachère . C'est encore une autre variante de la patrimonialisation en cours : la patrimonialisation partielle d'un site, qui le démembre, en un sens, tout en le réaffectant à une fonction culturelle contemporaine complètement séparée de l'histoire du lieu, ce qui est brillamment documenté dans l'exposition photographique de Jorge.

Author and Contributors

Jorge Ribalta is an artist, researcher, editor and independent curator.

Has made solo shows at, among others, galleries Zabriskie (New York and Paris), Casa sin Fin (Madrid) and angelsbarcelona (Barcelona). His recent solo exhibition *Monument Machine* was presented in 2015 at Museo Guerrero (Granada) and Fundación Helga de Alvear (Cáceres) and in 2016 at the Württembergischer Kunstverein (WKV), Stuttgart. Has participated in many group shows like *New Photography 10* (MOMA, New York, 1994), *Fragments* (MACBA, Barcelona, 1996) *Sets and Situations* (MOMA, New York, 2000), *Playgrounds* (Museum Reina Sofia, Madrid, 2014) and *The Beast and is the Sovereign* (MACBA and WKV, Barcelona and Stuttgart, 2015).

Curator of exhibitions (selection): *Universal Archive. The condition of the document and the modern photographic utopia* (MACBA, Barcelona, 2008), *A Hard, Merciless Light. The Worker-Photography Movement, 1926–1939* (Museum Reina Sofia, Madrid, 2011), and *Not Yet. On the Reinvention of Documentary and the Critique of Modernism* (Museum Reina Sofia, Madrid, 2015)

Editor of the books (selection): *Public Photographic Spaces. Propaganda Exhibitions from Pressa to Family of Man, 1928–1955* (MACBA, Barcelona, 2009), *The Worker Photography Movement, 1926–1939. Essays and Documents* (Museum Reina Sofia, Madrid, 2011) and *Not Yet. On the Reinvention of Documentary and the Critique of Modernism. Essays and Documents, 1972–1991* (Museum Reina Sofia, Madrid, 2015).

Muriel Enjalran is an art critic and a curator. Since 2015, she has been director of the CRP/ Centre régional de la photographie Hauts-de-France (formerly Nord-Pas-de-Calais), Douchy-les-Mines.

Jacques Lemière is professor of sociology and anthropology at Université de Lille. He is a major scholar on Portuguese cinema.

Hilde Van Gelder is professor of art history at the University of Leuven. She is director of the Lieven Gevaert Research Centre for Photography, Art and Visual Culture.

Colophon

Lieven Gevaert Research Centre for Photography, Art and Visual Culture
Arts Faculty KU Leuven
Blijde-Inkomststraat 21 box 3313
B-3000 Leuven
Belgium

Book author: Jorge Ribalta
Contributing authors to the panel discussion: Muriel Enjalran, Jacques Lemière, Jorge Ribalta and Hilde Van Gelder.
Translation into English: Anne Losq
Lay-out and cover design: DOGMA

ISBN 978 94 6270 159 5
D/2018/1869/55
NUR: 652

Published in collaboration with CRP/ Centre régional de la photographie Hauts-de-France, Württembergischer Kunstverein Stuttgart and Gallery Casa sin fin, Madrid.